Operator Theory
Advances and Applications
Vol. 71

Editor
I. Gohberg

Toeplitz Operators and Related Topics

The Harold Widom Anniversary Volume

Workshop on Toeplitz and Wiener-Hopf Operators,
Santa Cruz, California, September 20–22, 1992

Edited by

E.L. Basor
I. Gohberg

Springer Basel AG

Volume Editorial Office:

Raymond and Beverly Sackler Faculty of Exact Sciences
School of Mathematical Sciences
Tel Aviv University
69978 Tel Aviv
Israel

A CIP catalogue record for this book is available from the Library of Congress, Washington D.C., USA

Die Deutsche Bibliothek – CIP-Einheitsaufnahme
Toeplitz operators and related topics : the Harold Widom
anniversary volume ; workshop on Toeplitz and Wiener-Hopf
Operators, Santa Cruz, California, September 20–22, 1992 /
[vol. ed. office: Raymond and Beverly Sackler Faculty of Exact
Sciences, School of Mathematical Sciences, Tel Aviv
University]. Ed. by E.L. Basor ; I. Gohberg. – Basel ; Boston ;
Berlin : Birkhäuser, 1994
(Operator theory ; Vol. 71)
ISBN 978-3-0348-9672-6   ISBN 978-3-0348-8543-0 (eBook)
DOI 10.1007/978-3-0348-8543-0
NE: Basor, Estelle L. [Hrsg.]; Workshop on Toeplitz and Wiener-Hopf
Operators <1992, Santa Cruz, Calif.>; Bêt has-Sēfer le-Maddā'ê ham-
Matēmātîqā <Rāmat-Āvîv> / haf-Fāqûltā le-Maddā'îm
Medûyyaqîm'al Sēm Raymônd û-Beverly Saqler; GT

© 1994 Springer Basel AG
Originally published by Birkhäuser Verlag in 1994
Softcover reprint of the hardcover 1st edition 1994

Camera-ready copy prepared by the editors
Printed on acid-free paper produced from chlorine-free pulp
Cover design: Heinz Hiltbrunner, Basel

ISBN 978-3-0348-9672-6

9 8 7 6 5 4 3 2 1

Table of Contents

HAROLD WIDOM

Operator Theory:
Advances and Applications, Vol. 71
© 1994 Birkhäuser Verlag Basel/Switzerland

BIOGRAPHY OF H. WIDOM

Estelle L. Basor and Edward M. Landesman

A special meeting on Toeplitz and Wiener-Hopf operators was held in Santa Cruz, California on September 20-22, 1992, to celebrate the 60th birthday of Harold Widom and to recognize his contributions to the mathematical community. The meeting was sponsored jointly by the Mathematics Department of the University of California at Santa Cruz and the Mathematical Sciences Research Institute in Berkeley. The authors of this essay both spoke about the life and work of Harold Widom at the conference, and we are repeating many of our remarks for the present readers.

The career of Harold Widom has been distinguished by his research, teaching, and service to the mathematical profession. He was born September 23, 1932, in Newark, New Jersey during the heart of the Depression. His parents were born in eastern Europe, his father in Kamentz-Podolskii in the Ukraine and his mother in Mielic, Poland. Both of his parents came to the United States in 1914, when his mother was 15 years old and his father 22. They met in New York and were married there in 1924.

Harold grew up in Brooklyn, New York. He went to Stuyvesant High School where he was captain of the math team. Coincidentally, the captain of the rival team at the Bronx High School of Science was Henry Landau, who also attended the conference and has been a long-time friend and colleague of Harold's. Also attending the conference was Harold's brother Benjamin, who is five years older and is Professor of Chemistry at Cornell University.

After high school, he attended the City College of New York and the University of Chicago, where at the age of twenty he received a Master's Degree. Three years later, in 1955, he obtained his Ph.D. in mathematics, having written a dissertation in functional analysis, under the direction of Irving Kaplansky. He began his academic career as an instructor at Cornell University where he rose through the ranks to become full professor in 1965. At Cornell, his interests focused on classical analysis and operator theory and he proved many of the early beautiful theorems about Toeplitz operators. In particular, he gave necessary and sufficient conditions for a Toeplitz operator to be invertible, proved an index theorem for Toeplitz operators with piecewise continuous symbol, and proved the still somewhat mysterious result that the spectrum of a Toeplitz operator is connected. This analysis was characterized by a special style -- very deep theorems with difficult proofs, motivated and described in a most elegant, clear way.

In 1968, after visiting Stanford University for one year, a rumor surfaced that Harold enjoyed California weather. Happily for the University of California at Santa Cruz the mathematics department was able to convince him to join its faculty in the fall of 1968. At Santa Cruz, his research evolved from the study of Szegö Limit Theorems. This theorem gave the two first terms in the asymptotic expansion of the determinants of finite Toeplitz matrices. The proofs of the Szegö theorems were difficult, indirect, and the value of the constant term did not follow intuitively from anything else.

Harold changed this state of affairs in 1976 when he gave a beautiful operator theory proof of the strong Szegö Limit Theorem, showed that the constant term was simply an infinite determinant, and extended the result to the block matrix case. He also extended the original Szegö result to Wiener-Hopf operators, Toeplitz operators with singular symbols, and, in the most general setting, to pseudodifferentail operators.

For pseudodifferential operators on manifolds he developed a complete symbolic calculus, which allowed him to find a complete Szegö expansion for n-dimensional convolution operators on general domains. He also showed how the Szegö expansion and analogues of the heat expansion of the Laplace operator could be computed with one common pseudodifferential operator approach. The is the subject of his book *Asymptotic Expansions for Pseudodifferential Operators on Bounded Domains*, which appeared in Lecture Notes in Mathematics. Most recently, his interests have turned to the study of random matrices and their relationships to Toeplitz and Wiener-Hopf operators.

Throughout his career, Harold has been known as a superb and inspiring teacher. The same "Widom" style so evident in his research is evident also in his teaching. His lectures are polished, clear, and beautifully motivated. His ability to uncover the heart of a subject and to make complicated details seem simple has attracted many students to analysis. He has directed the dissertations of seven Ph.D. students. His first student, Lydia Luquet, was a student at Cornell. The six others include the first author, Ray Roccaforte, Xiang Fu, Shuxian Lou, Richard Libby, and Bobette Thorsen.

On a personal note, the first author was a student in an undergraduate real analysis course taught by Harold in the fall and winter of 1968-69 at Santa Cruz. For her, those courses were the most delightful in her undergraduate career. Harold managed to teach analysis in a very rigorous fashion, yet allowed the intuition and beauty to come through. His love and respect for mathematics is always apparent and has influenced many a student and colleague.

Harold has also worked in many ways for the mathematics community. He has been on the editorial boards for the Journal of Integral Equations and Operator Theory and the journal for Asymptotic Analysis and an organizer for several conferences at Oberwolfach. In addition to *Asymptotic Expansions for Pseudodifferential Operators on Bounded Domains*, he has written two books in the Van Nostrand Mathematical Studies series, *Lectures on Integral Equations* and *Lectures on Measure and Integration*.

On a local level, he has served on numerous department committees and was also chairman of the Santa Cruz Mathematics Department for 3 years. His concern for the mathematical community is not restricted to university mathematics. Last year he wrote all the problems for the local Santa Cruz high school mathematics contest.

Numerous awards and grants have been bestowed upon Harold. In 1951, he was a Putnam fellow in the eleventh Putnam examination. He has received an NSF Postdoctoral Fellowship, several NSF research grants, a Sloan Fellowship and was twice the recipient of a Guggenheim Fellowship.

However, his most enduring mark on the mathematical community is that we, his colleagues and students, try to emulate him. We hope our mathematics is as beautiful and elegant. We hope that we treat others with the same dignity and respect that he does, and we hope we work in the same unassuming, dedicated way. We present this volume as a celebration of his work.

Mathematics Department
California Polytechnic State University
San Luis Obispo, CA 93407

Department of Mathematics
University of California at Santa Cruz
Santa Cruz, CA 95064

Operator Theory:
Advances and Applications, Vol. 71
© 1994 Birkhäuser Verlag Basel/Switzerland

TO HAROLD WIDOM ON HIS 60TH BIRTHDAY

Irving Kaplansky

September 23, 1992 was the 60th birthday of Harold Widom. During the three preceding days, at the Santa Cruz campus of the University of California, there was a conference in his honor on the topic: Wiener-Hopf and Toeplitz Operators. It was a pleasant and fruitful get-together of his colleagues and students, as I can personally attest.

In 1946 Marshall Stone left Harvard to accept the chairmanhsip of the Department of Mathematics at th University of Chicago. There followed quickly a series of stellar appointments that raised the department to a very high level. (I can say this without being self-serving; John Kelley and I were the last appointments made before the "Stone Age".) It was an exciting time to be at Chicago. But it was not only the faculty that created the excitement – a stream of superb students arrived. I was lucky enough to attract my fair share, and that included Harold. His thesis was on AW^*-algebras (I shall say more about this shortly). His bibliography shows three fine papers on the topic and then shifts. (With the shift, his output moved to a different part of Mathematical Reviews). I understand that the shift can be attributed to the influence of Marc Kac at Cornell and one could not ask for a better source of inspiration. I am proud and happy about what Harold added to the theory of AW^*-algebras. and equally proud and happy about what he has accomplished since then.

Now let me redeem the promise to say a little about AW^*-algebras. The story begins with the launching in 1936 by Murray and von Neumann of a highly original study of weakly closed self-adjoint algebras of operators on a Hilbert space. They called them "rings of operators". The algebras were renamed twice. The French school likes to name mathematical objects for people and called them von Neumann algebras. Segal proposed the name W^*-algebras, as a companion for his C^*-algebras. Here C stands for uniformly closed, W for weakly closed.

The verdict in this terminology battle is in and it is mixed. Von Neumann algebras have won out, but the French attempt to replace C^*-algebras by Gelfand-Neumark algebras failed.

For C^*-algebras Gelfand and Neumark had given a neat set of intrinsic axioms that attracted a lot of attention. Presumably, it was the desire to do something similar for von Neumann algebras that motivated Hille to suggest to Rickart the project that matured into AW^*-algebras. It is not quite historically accurate, but I shall say that the idea was to add to the axioms for C^*-algebras one little purely algebraic assumption: that the right annihilator of any subset can be generated by an idempotent. This does not characterize von Neumann algebras, but it comes close. I think it is a fair statement that the work on AW^*-algebras helped to illuminate and enrich the theory of von Neumann algebras, and more generally C^*-algebras. Harold's work was an important ingredient.

André Weil once thanked me for naming a class of algebras for him. (Of course, he was well aware that the letter A simply stands for "abstract".)

The subject promises to run through nine lives. In 1985 Gert Pedersen gave a talk in Berkeley entitled: "AW^*-algebras – the subject that refused to die".

I have assembled a short appended bibliography, beginning with [1], the definitive book (Berberian was a Chicago student who was contemporary with Harold) and including Harold's three papers. The remaining items record advances that I find particularly striking. I rashly conjectured that the difference between von Neumann algebras and AW^*-algebras resides entirely in the center; this was defeated in [2] and [5]. (I am indebted to John Wright for the information that, though the examples in these two papers appear to be different, they are now known to be isomorphic; see [10]. Also noteworthy is [9], where it is shown that "wild" AW^*-algebras are by no means rare. I made a happier conjecture that the cardinal number attached to a homogeneous type I AW^*-algebra need not be unique; this was confirmed in [4]. Paper [3] proved the theorem stated in its title, thus bringing AW^*-algebras in line with von Neumann algebras on this front. Still holding out is the challenge to settle whether the dimension function on an AW^* factor of type

II_1 extend to a trace; on this I am cautiously not offering any conjecture.

In closing, I take this public opportunity to wish Harold and Linda many more years to continue the wonderful personal and professional inspiration they have given to so many people.

REFERENCES

1. Sterling Berberian, Baer *-Rings, Springer, 1972.

2. John Dyer, Concerning AW^*-algebras, Notices Amer. Math. Soc. 17(1970), 788; abstract no. 677-47-5.

3. Dorte Olsen, Derivations of AW^*-algebras are inner, Pac. J. of Math. 53(1974), 555-561

4. Masanao Ozawa, Nonuniqueness of the cardinality attached to homogeneous AW^*-algebras, Proc. Amer. Math. Soc. 93(1985), 681-684.

5. Osamu Takenouchi, A non-W^*, AW^*-factor, pp. 131-134 in C^*-algebras and applications to Physics, Springer Lecture Notes 650, 1978.

6. Harold Widom, Embedding in algebras of type I, Duke Math. J. 23(1956), 309-324.

7. _____, Approximately finite algebras, Trans. Amer. Math. Soc. 83(1956), 170-178.

8. _____, Nonisomorphic approximately finite factors, Proc. Amer. Math. Soc. 8(1957), 537-540.

9. J.D. Maitland Wright, Wild AW^*-factors and Kaplansky-Rickart algebras, J. London Math. Soc. 13(1976), 83-89.

10. Dennis Sullivan, B. Weiss, and J.D. Maitland Wright, Generic dynamics and monotone complete C^*-algebras, Trans. Amer. Math. Soc. 295(1986), 795-809.

Mathematical Sciences Research Institute

1000 Centennial Drive

Berkeley, CA 94720-0001, U.S.A.

Operator Theory:
Advances and Applications, Vol. 71
© 1994 Birkhäuser Verlag Basel/Switzerland

Harold Widom's Publications

Papers

1. *Embedding in algebras of type I*, Duke Math. J. 23 (1956) 309-324.

2. *Approximately finite algebras*, Trans. Amer. Math. Soc. 83 (1956) 170-178.

3. *Nonisomorphic approximately finite factors*, Proc. Amer. Math. Soc. 8 (1957) 537-540.

4. *Equations of Wiener-Hopf type*, Ill. J. Math. 2 (1958) 261-270.

5. *On the eigenvalues of certain Hermitian operators*, Trans. Amer. Math. Soc. 88 (1958) 491-522.

6. (with H. Pollard) *Inversion of an integral transform*, Proc. Amer. Math. Soc. 9 (1958) 598-602.

7. (with A. Calderon and F. Spitzer) *Inversion of Toeplitz matrices*, Ill. J. Math. 3 (1959) 490-498.

8. *Inversion of Toeplitz matrices II*, Ill. J. Math. 4 (1960) 88-99.

9. *A theorem on translation kernels in n dimensions*, Trans. Amer. Math. Soc. 94 (1960) 170-180.

10. (with M. Rosenblum) *Two extremal problems*, Pac. J. Math. 10 (1960) 1409-1418.

11. *Singular integral equations in L_p*, Trans. Amer. Math. Soc. 97 (1960) 131-160.

12. *Stable processes and integral equations*, Trans. Amer. Math. Soc. 98 (1961) 430-449.

13. (with F. Spitzer) *The circumference of a convex polygon*, Proc. Amer. Math. Soc. 12 (1961) 506-509.

14. *A two-sided absorption problem*, Proc. Amer. Math. Soc. 12 (1961) 862-869.

15. *Extreme eigenvalues of translation kernels*, Trans. Amer. Math. Soc. 100 (1961) 252-262.

16. *Extreme eigenvalues of N-dimensional convolution operators*, Trans. Amer. Math. Soc. 106 (1963) 391-414.

17. *Rapidly increasing kernels*, Proc. Amer. Math. Soc. 14 (1963) 501-506.

18. *Asymptotic behavior of the eigenvalues of certain integral equations*, Trans. Amer. Math. Soc. 109 (1963) 278-295.

19. *On the spectrum of a Toeplitz operator*, Pac. J. Math. 14 (1964) 365-375.

20. *Asymptotic behavior of the eigenvalues of certain integral equations II*, Arch. Rat. Mech. Anal. 17 (1964) 215-229.

21. *Toeplitz matrices*, in "Studies in Real and Complex Analysis" (Ed. I. I. Hirschman), Studies in Math. 3 (1965) 179-201, Math. Assoc. Amer.

22. *Hankel matrices*, Trans. Amer. Math. Soc. 121 (1966) 1-35.

23. *Toeplitz operators on H_p*, Pac. J. Math. 19 (1966) 573-582.

24. (with H. Wilf) *Small eigenvalues of large Hankel matrices*, Proc. Amer. Math. Soc. 17 (1966) 338-344.

25. *Polynomials associated with measures in the complex plane*, J. Math. Mech. 16 (1967) 997-1014.

26. *Norm inequalities for entire functions of exponential type*, in Proc. Conf. on Orthogonal Expansions and their Continuous Analogues (Southern Illinois University, Edwardsville, April, 1967) 143-165.

27. *Extremal polynomials associated with a system of curves in the complex plane*, Adv. Math. 3 (1969) 127-232.

28. *On some extremal quantities associated with unbounded sets in the plane*, J. Math. Mech. 19 (1969) 439-450.

29. *An inequality for rational functions*, Proc. Amer. Math. Soc. 24 (1970) 415-416.

30. (with R.G. Douglas) *Toeplitz operators with locally sectorial symbols*, Ind. U. Math. J. 20 (1970) 385-388.

31. *The maximum principle for multiple-valued analytic functions*, Acta Math. 126 (1971) 63-82.

32. *H_p sections of vector bundles over Riemann surfaces*, Ann. Math. 94 (1971) 304-324.

33. *The strong Szegö limit theorem for circular arcs*, Ind. U. Math J. 21 (1971) 277-283.

34. *Rational approximation and n-dimensional diameter*, J. Approx. Th. 5 (1972) 343-361.

35. *Toeplitz determinants with singular generating functions*, Amer. J. Math. 95 (1973) 333-383.

36. *Asymptotic behavior of block Toeplitz matrices and determinants*, Adv. Math. 13 (1974) 284-322.

37. *Asymptotic inversion of convolution operators*, Publ. Math. I. H. E. S. 44 (1975) 191-240.

38. *On the limit of block Toeplitz determinants*, Proc. Amer. Math. Soc. 50 (1975) 167-173.

39. *Perturbing Fredholm operators to obtain invertible operators*, J. Funct. Anal. 20 (1975) 26-31.

40. *Asymptotic behavior of block Toeplitz matrices and determinants II*, Adv. Math. 21, (1976) pp. 1-29.

41. *Asymptotic expansions of determinants for families of trace class operators*, Ind. U. Math. J. 27 (1978) 449-478.

42. *Eigenvalue distribution theorems in certain homogeneous spaces*, J. Funct. Anal. 32 (1979) 139-147.

43. *The Laplace operator with potential on the 2-sphere*, Adv. Math. 31 (1979) 63-66.

44. *Szegö's theorem and a complete symbolic calculus for pseudodifferential operators*, in "Seminar on Singularities" (Ed L. Hörmander), Ann. of Math. Studies 91 (1979) 261-283, Princeton Univ. Press.

45. *Families of pseudodifferential operators*, in "Topics in Functional Analysis" (Ed. I. Gohberg and M. Kac) (1978) 345-395, Academic Press.

46. *A complete symbolic calculus for pseudodifferential operators*, Bull. Sciences Math. 104 (1980) 19-63.

47. *Spectral asymptotics of hypersurfaces*, Int. Eqs. Oper. Th. 1 (1978) 415-443.

48. *Compressions to spectral subspaces of the Laplacian*, Proc. Symp. Pure Math. 36 (1980) 319-323, Amer. Math. Soc.

49. (with H. J. Landau) *Eigenvalue distribution of time and frequency limiting*, J. Math. Anal. 77 (1980) 469-481.

50. *Szegö's limit theorem: the higher-dimensional matrix case*, J. Funct. Anal. 39 (1980) 182-198.

51. *Szegö's theorem and pseudodifferential operators*, Proc. Canadian Math. Soc. Seminar on Harmonic Analysis (1980) 165-176.

52. *Second-order spectral asymptotics of certain integral operators*, in "Probability, Statistical Mechanics, and Number Theory" (Ed. G.-C. Rota), (1986) 63-80, Academic Press.

53. *On a class of integral operators with discontinuous symbol*, Oper. Th.: Adv. Appl. 4 (1982) 477-500.

54. *A trace formula for Wiener-Hopf operators*, J. Oper. Th. 8 (1982) 279-298.

55. (with E. Basor) *Toeplitz and Wiener-Hopf determinants with piecewise continuous symbols*, J. Funct. Anal. 50 (1983) 387-413.

56. *Spectral asymptotic expansions for pseudodifferential operators*, Proc. 8th Conf. on Oper. Th. (Timisoara and Herculane, 1983), Birkhäuser-Verlag, 1984.

57. *Asymptotic expansions for pseudodifferential operators on bounded domains.* Lecture Notes in Math. 1152 (1985), Springer-Verlag.

58. *More about pseudodifferential operators on bounded domains*, Proc. 9th Conf. on Oper. Th. (Timisoara and Herculane, 1984), Birkhäuser-Verlag, 1986.

59. *Trace formulas for Wiener-Hopf operators*, Proc. 10th Conf. on Oper. Th. (Bucharest 1985), Birkhäuser-Verlag, 1987.

60. *The heat expansion for systems of integral equations*, Oper. Th.: Adv. Appl. 35 (1988) 495-521.

61. *On an inequality of Osgood, Phillips, and Sarnak*, Proc. Amer. Math. Soc. 12 (1988) 773-774.

62. *The heat expansion for systems of integral equations II*, Asymp. Anal. 1 (1988) 95-103.

63. *On Weiner-Hopf determinants*, Oper. Th.: Adv. Appl. 41 (1989) 519-543.

64. *Integral operators on a half-space with discontinuous symbol*, J. Funct. Anal. 88 (1990) 166-193.

65. *On the sigular values of Toeplitz matrices*, Zeit. f. Anal. Anw. 8 (1989) 221-229.

66. *Szegö expansions for operators with smooth or nonsmooth symbol*, in Proc. Symp. Pure Math. Soc. 51 (1990) 599-615, Amer. Math. Soc.

67. *Eigenvalue distribution of nonselfadjoint Toeplitz matrices and the asymptotics of Toeplitz determinants in the case of nonvanishing index*, Oper. Th.: Adv. Appl. 48 (1990) 387-421.

68. (with B. Widom) *Model for line tension in three-phase equilibrium*, Physica A 173 (1991) 72-110.

69. *Asymptotic expansions and stationary phase for operators with discontinuous symbol*, Asymp. Anal. 4 (1994) 35-63.

70. *Symbols and asymptotic expansions*, Oper. Th.: Adv. Appl. 58 (1992) 189-210.

71. (with E. Basor and C. A. Tracy) *Asymptotics of level- spacing distributions for random matrices*, Phys. Rev. Letts 69 (1992) 5-8.

72. *The asymptotics of a continuous analogue of orthogonal polynomials.* J. Approx. Th. 76 (1994).

73. (with C. A. Tracy) *Introduction to random matrices* in "Geometric and Quantum Aspects of Integrable Systems" (Ed. G. F. Helminck), Lecture Notes in Physics 424 (1993) 103-130, Springer-Verlag.

74. (with C. A. Tracy) *Level-spacing distributions and the Airy kernel,* Phys. Letts B 305 (1993) 115-118.

75. (with C. A. Tracy) *Level-spacing distributions and the Airy kernel,* Comm. Math. Phys. 159 (1994) 151-174.

76. (with J. Harnad and C.A. Tracy) *Hamiltonian structure of equations appearing in random matrices,* in "Low-Dimensional Topology and Quantum Field Theory" (Ed. H. Osborn), NATO ASI Series B 314 (1993) 231-245, Plenum Press.

77. (with A. Böttcher and B. Silbermann) *A continuous analogue of the Fisher- Hartwig formula for piecewise continuous symbols,* J. Func. Anal. 120 (1994).

78. (with A. Böttcher) *Two remarks on spectral approximation for Wiener-Hopf operators,* J. Int. Eqs. Appl.

79. (with C. A. Tracy) *Fredholm determinants, differential equations and matrix models,* Comm. Math. Phys.

80. (with A. Böttcher and B. Silbermann) *Determinants of truncated Wiener-Hopf operators with Hilbert-Schmidt kernels and piecewise continuous symbols,* Archiv d. Math.

Books

1. *Lectures on Integral Equations,* Van Nostrand Mathematical Studies, No. 17, New York, 1969.

2. *Lectures on Measure and Integration,* Van Nostrand Mathematical Studies, No. 20, New York, 1979.

Operator Theory:
Advances and Applications, Vol. 71
© 1994 Birkhäuser Verlag Basel/Switzerland

EIGENVALUE DISTRIBUTION FOR NONSELFADJOINT TOEPLITZ MATRICES[1]

Harold Widom

In this expository article we discuss the question of the limiting distribution as $n \to \infty$ of the the the eigenvalues of $n \times n$ Toeplitz matrices, with emphasis on the determination of the limiting set and the limiting measure (if these exist). In the selfadjoint case the limiting set is the interval between the essential infimum and the essential supremum of the symbol, and the limiting measure is the canonical measure induced by the symbol. Theorems of Schmidt-Spitzer and Hirschman, which determine these when the symbol is a Laurent polynomial, are discussed. They are quite different from what they are in the selfadjopint case. A conjecture is presented (with some evidence given) that, nevertheless, the limiting measure is "in general" the one induced by the symbol.

Although there is a vast literature on the subject of eigenvalue distribution for selfadjoint operators of various kinds, results for nonselfadjoint operators are sparse. In this expository article we describe some of these in the context of finite Toeplitz matrices $(c_{i-j})_{i,j=0,\ldots,n-1}$. We are concerned with the limiting distribution of the eigenvalues of these matrices as $n \to \infty$, and will always assume that the c_k are the Fourier coefficients $\hat{\phi}(k)$ of a bounded function ϕ defined on the unit circle. As usual we denote the $n \times n$ matrix associated with ϕ by $T_n(\phi)$, and we denote its eigenvalues, repeated according to (algebraic) multiplicity, by $\lambda_{i,n}$ $(i = 1, \ldots, n)$.

There are two kinds of questions one might ask about the limiting distribution. The first is how the eigenvalues are distributed "on the average". More precisely, given a set E, does the limit

$$\lim_{n \to \infty} \frac{\#\{i : \lambda_{i,n} \in E\}}{n} \tag{1}$$

exist and, if it does, what is it? The second is the question of what the "limiting set" of the eigenvalues is. There are two possible definitions for this, which might be called the strong and weak limiting sets, given respectively by

$$\Lambda_s := \{\lambda : \lambda_{i_n,n} \to \lambda \text{ for some sequence } i_n \to \infty\},$$

$$\Lambda_w := \{\lambda : \lambda_{i_k,n_k} \to \lambda \text{ for sequences } i_k, n_k \to \infty\}.$$

---

[1] Toeplitz Lecture presented at Tel Aviv University, March, 1993

The answers to both of these questions are known in the selfadjoint case, when ϕ is real-valued, and we give them here.

The average distribution is described by the classical theorem of Szegö, which states that if f is any continuous function defined on $\mathbb{R}$ then

$$\lim_{n \to \infty} \frac{1}{n} \sum_{i=1}^{n} f(\lambda_{i,n}) = \frac{1}{2\pi} \int_{-\pi}^{\pi} f(\phi(e^{i\theta}))\, d\theta. \tag{2}$$

If one takes for f here the characteristic function of E then one finds that the limit (1) equals $\frac{1}{2\pi} m(\phi^{-1}(E))$, where "m" denotes Lebesgue measure on the circle. Of course (2) does not apply to this characteristic function, since it isn't continuous, but the resulting relation can be shown to hold if $f^{-1}(\partial E)$ has measure zero. And it shows that we can interpret (2) to mean that the eigenvalues of $T_n(\phi)$ are distributed, in the limit, as the values of ϕ.

Another interpretation of (2) is this. Denote by $\mu_{n,\phi}$ the measure which assigns each eigenvalue $\lambda_{i,n}$ measure $1/n$ and by μ_ϕ the measure given by

$$\mu_\phi(E) = \frac{1}{2\pi} m(\phi^{-1}(E)).$$

Then (2) is equivalent to the statement that

$$\mu_{n,\phi} \to \mu_\phi \text{ weakly as } n \to \infty.$$

We say that in this case the eigenvalues are "canonically distributed".

As for the limiting set(s) in this selfadjoint case, the answer is particularly simple:

$$\Lambda_s = \Lambda_w = [\text{ess inf } \phi, \text{ ess sup } \phi]. \tag{3}$$

Here is the proof. Write Λ for the set on the right. Then Λ contains the numerical range of $T_n(\phi)$ for all n and so, clearly, $\Lambda_w \subset \Lambda$. Next, assume $\lambda \notin \Lambda_s$. Then for some sequence of n's tending to ∞ and some $\delta > 0$ no $\lambda_{i,n}$ belongs to the interval $(\lambda - \delta, \lambda + \delta)$. Since $T_n(\phi)$ is selfadjoint we have, for such n, the inequality

$$\| (T_n(\phi) - \lambda I)\, x \| \geq \delta \| x \|$$

for all vectors x. Hence, letting $n \to \infty$, we obtain

$$\| (T(\phi) - \lambda I)\, x \| \geq \delta \| x \|$$

where $T(\phi)$ is the semi-infinite Toeplitz matrix with symbol ϕ. Since $T(\phi)$ is selfadjoint, this inequality implies that $T(\phi) - \lambda$ is invertible. But it's well-known that the spectrum of $T(\phi)$ is precisely the set Λ. Hence $\lambda \notin \Lambda_s$ implies $\lambda \notin \Lambda$. Thus we have shown

$$\Lambda_w \subset \Lambda \subset \Lambda_s,$$

which implies (3).

We shall also give the idea of the proof of (2) to see what we might get from it in the nonselfadjoint case. Given ϕ_1 and ϕ_2 one has

$$T_n(\phi_1)\, T_n(\phi_2) \approx T_n(\phi_1\, \phi_2),$$

where "$\approx$" means "roughly equal to". (This is a reflection of the fact that the sequence of Fourier coefficients of a product of two functions is the convolution of the two sequences of Fourier coefficients.) By induction we deduce

$$T_n(\phi)^k \approx T_n(\phi^k),$$

and so the traces are roughly equal. Since, for any ψ,

$$\text{tr } T_n(\psi) = n \, \hat{\psi}(0) = \frac{n}{2\pi} \int \psi(e^{i\theta}) \, d\theta,$$

the above yields

$$\text{tr } T_n(\phi)^k \approx \frac{n}{2\pi} \int \phi(e^{i\theta})^k \, d\theta.$$

This (when done precisely) yields (2) for $f(\lambda) = \lambda^k$ for any k and so for polynomial f. Since both $\lambda_{i,n}$ and $\phi(e^{i\theta})$ are real, and since the polynomials in λ are uniformly dense in the space of continuous functions on a compact interval in $\mathbb{R}$, (2) follows easily for general continuous f.

Clearly this argument fails for nonselfadjoint matrices since polynomials in λ are not uniformly dense in the space of continuous functions on a compact subset of $\mathbb{C}$. But it does give a result for analytic f. The convex hull, Σ, of the essential range of ϕ is simply-connected and all $\lambda_{i,n} \in \Sigma$. Therefore the same approximation argument shows that (2) holds if f is analytic on (a neighborhood of) Σ. In particular, taking $f = \log$ gives an asymptotic formula for the Toeplitz determinants

$$D_n(\phi) := \det T_n(\phi)$$

under the assumption that $0 \notin \Sigma$. It reads

$$\lim_{n \to \infty} \frac{1}{n} \log D_n(\phi) = \frac{1}{2\pi} \int \log \phi(e^{i\theta}) \, d\theta,$$

or equivalently

$$\lim_{n \to \infty} D_n(\phi)^{1/n} = G(\phi) \tag{4}$$

where

$$G(\phi) := \exp \{\frac{1}{2\pi} \int \log \phi(e^{i\theta}) \, d\theta\}.$$

This holds for other ϕ as well. For example it holds if ϕ is continuous, nonzero, and satisfies

$$i(\phi) := \frac{1}{2\pi} \arg \phi(e^{i\theta}) \mid_{-\pi}^{\pi} = 0.$$

The relation (4) is fundamental in the investigation of eigenvalues of nonselfadjoint Toeplitz matrices. For example, it follows immediately from a uniform version of the last-mentioned result that for continuous ϕ we have

$$\Lambda_w \subset \text{range } \phi \cup \{\lambda \notin \text{range } \phi : i(\phi - \lambda) \neq 0\}.$$

This is just because λ is not an eigenvalue of $T_n(\phi)$ precisely when $D_n(\phi - \lambda) \neq 0$.

The eigenvalues of $T_n(\phi)$ are not always caconically distributed. For example if ϕ extends analytically into $|z| < 1$ or $1 < |z| \leq \infty$ then $T_n(\phi)$ is triangular and its sole eigenvalue is $\hat{\phi}(0)$ (of multiplicity n).

The first nontrivial results about the limiting eigenvalue distribution in the non-selfadjoint case are due to Schmidt and Spitzer [10], who considered the case of a Laurent polynomial

$$\phi(z) = \sum_{k=-p}^{q} c_k z^k \qquad (p, q > 0, c_{-p} \neq 0, c_q \neq 0). \tag{5}$$

They observed that if one defines $\phi_r(z) := \phi(rz)$ then the matrices $T_n(\phi)$ and $T_n(\phi_r)$ are similar via the diagonal matrix $diag(1, r, \ldots, r^{n-1})$ and so have the same eigenvalues. The rational function $\phi(z) - \lambda$ has $p + q$ zeros in the complex plane; denote them by $z_i(\lambda)$ and assume they are ordered so that

$$|z_1(\lambda)| \leq |z_2(\lambda)| \leq \ldots \leq |z_{p+q}(\lambda)|.$$

If $|z_p(\lambda)| < |z_{p+1}(\lambda)|$ choose r so that

$$|z_p(\lambda)| < r < |z_{p+1}(\lambda)|. \tag{6}$$

With this r we have $\phi_r(e^{i\theta}) - \lambda \neq 0$, and $i(\phi_r - \lambda) = 0$ by the argument principle. Hence some neighborhood of such a λ cannot contain any eigenvalues if n is large enough. This implies that if we set

$$\Lambda := \{\lambda : |z_p(\lambda)| = |z_{p+1}(\lambda)|\}$$

then $\Lambda_w \subset \Lambda$. Schmidt and Spitzer showed that this set Λ is a union of analytic curves and arcs, and that in fact

$$\Lambda_w = \Lambda_s = \Lambda. \tag{7}$$

Hirschman [7] completed this result by finding the limiting distribution of the eigenvalues; in other words, he determined the weak limit of the measures $\mu_{n,\phi}$ for this case. Although Hirschman did not do this by exactly the method we now describe, he did introduce potential-theoretic ideas to the problem.

It is a fact from potential theory that for any compactly supported finite Borel measure μ one has

$$\frac{1}{2\pi} \Delta \int \log|z - \lambda| \, d\mu(z) = \mu \tag{8}$$

where Δ denotes the distributional Laplacian with respect to the λ variable. In the special case $\mu = \mu_{n,\phi}$ this identity becomes

$$\frac{1}{2\pi} \Delta \log |D_n(\phi - \lambda)|^{1/n} = \mu_{n,\phi}. \tag{9}$$

From this fact one easily derives the following [12].

PROPOSITION 1. *Assume there is a function $g(\lambda)$, positive a.e., such that*

$$|D_n(\phi - \lambda)|^{1/n} \to g(\lambda) \quad a.e..$$

Then $\log g$ is locally integrable, $\Delta \log g$ is a measure, and $\mu_{n,\phi}$ converges weakly to $\frac{1}{2\pi} \Delta \log g$ as $n \to \infty$.

Let us deduce Hirschman's result for the Laurent polynomials (5). In the neighborhood of any point not in the set Λ we can find an r so that (6) holds for λ in that neighborhood. Then (4) holds with ϕ replaced by $\phi_r - \lambda$ and so for such λ

$$\lim_{n \to \infty} |D_n(\phi - \lambda)^{1/n}| = |G(\phi_r - \lambda)| = \exp\{\frac{1}{2\pi} \int \log |\phi_r(e^{i\theta}) - \lambda| \, d\theta\}.$$

The function in brackets, defined for $\lambda \notin \Lambda$, is harmonic in the complement of Λ (a set of measure 0) and extends continuously up to Λ with a jump in its normal derivative. (Hirschman showed this.) It follows that the limiting measure exists, is supported on Λ, and in fact equals $\frac{1}{2\pi}$ times arc length measure on Λ times the jump in this normal derivative. This was Hirschman's result. (Observe that one consequence is that $\Lambda \subset \Lambda_s$ and this completes the demonstration of (7).)

Day [5] extended the results of Schmidt-Spitzer and Hirschman to the case of arbitrary ϕ and, although complicated, the situation is well understood in this case. Of course, it is only rarely that we have canonical eigenvalue distribution.

But there is evidence that the following assertion has some merit. *If ϕ is a "normal" function which, however, cannot be analytically continued to an annulus $r < |z| < 1$ or $1 < |z| < r$, then the eigenvalues of $T_n(\phi)$ are canonically distributed.* Although it is hopeless to try to prove such a statement we shall present some evidence in its favor.

Identity (8) with $\mu = \mu_\phi$ can be restated as

$$\frac{1}{2\pi} \Delta \frac{1}{2\pi} \int \log |\phi(e^{i\theta}) - \lambda| \, d\theta = \mu_\phi$$

and so Proposition 1 implies

PROPOSITION 2. *If the essential range of ϕ has measure zero and*

$$\lim_{n \to \infty} |D_n(\phi - \lambda)|^{1/n} \to G(|\phi - \lambda|) \tag{10}$$

a.e. then the eigenvalues of $T_n(\phi)$ are canonically distributed.

Let us consider first the case of a continuous function ϕ. We know that (10) holds if $i(\phi - \lambda) = 0$, but what if $i(\phi - \lambda) \neq 0$? We do the following. Suppose $\lambda \notin$ range ϕ but $i(\phi - \lambda) = 1$ and define ϕ_0 by

$$\phi_0(z) := z^{-1}(\phi(z) - \lambda).$$

Then $T_{n-1}(\phi - \lambda)$ is the $n - 1, 0$ minor of the matrix $T_n(\phi_0)$ and so Cramer's rule gives

$$\left(T_n(\phi_0)^{-1}\right)_{0,\,n-1} = (-1)^{n-1} \frac{D_{n-1}(\phi - \lambda)}{D_n(\phi_0)}.$$

More generally, if $i(\phi - \lambda) = m > 0$ and one sets $\phi_0(z) := z^{-m}(\phi(z) - \lambda)$ then

$$D_{n-m}(\phi - \lambda) = \pm D_n(\phi_0) \det \left(T_n(\phi_0)^{-1}_{i,\,n-m+j}\right) \tag{11}$$

where in the determinant on the right $i, j = 0, \ldots, m - 1$. (A similar formula holds when $i(\phi - \lambda) < 0$.)

Now since ϕ_0 is continuous with index zero the asymptotics of $D_n(\phi_0)$ are given by (4). Since $|\phi_0| = |\phi - \lambda|$, (10) will follow if the determinant on the right side of (11) is not exponentially small as $n \to \infty$. (It cannot be exponentially large; in fact, it's bounded.) Using this approach we considered the case where ϕ is in addition piecewise C^∞ but not C^∞ [12]. Canonical eigenvalue distribution was demonstrated in case ϕ has only one singularity (point in no neighborhood of which ϕ is C^∞) and in other cases as well, although a proof in the general case remains elusive. A consequence of the proof was that

$$\Lambda_s = \Lambda_w = \text{range}\,\phi \tag{12}$$

in the case of one singularity and in some, but very likely not all, of the other cases as well.

To add further credence to the conjecture, it was shown that if $\phi(e^{i\theta})$ describes a simple smooth curve C then the following hold:

(i) If a weak limit of some subsequence of the measures $\mu_{n,\phi}$ has support inside C then ϕ continues to an annulus, and if ϕ does so continue then all such limiting measures have support inside C.

(ii) If the support of any limiting measure is a subset of C then that limiting measure must be μ_ϕ.

What was required in this investigation (not the last statements–those used only (4) for continuous ϕ with index zero) was very fine knowledge of the asymptotics of the entries of the inverse matrix $T_n(\phi_0)^{-1}$. That is why the situation is so much more difficult when ϕ is discontinuous. There is an old conjecture of Fisher and Hartwig [6], on aspects of which people are still working to this day, which implies that if ϕ is nonzero and smooth except for jump discontinuities then

$$|D_n(\phi)| \sim c\,G(|\phi|)^n\,n^{-e} \tag{13}$$

where e is an exponent depending on the jumps of ϕ, and c is a nonzero constant. This, when true, can serve as a replacement for (4), and then (11) might be used to extend the result to other ϕ's.

What is known, first, is that (13) holds in various cases where ϕ satisfies a "small index" condition (replacing the zero index condition under which (4) holds). To be precise, assume that ϕ has the form

$$\phi(e^{i\theta}) = \psi(e^{i\theta}) \prod_j e^{-i\beta_j(\theta-\theta_j)}$$

where ψ is smooth, nonzero, and has index zero, and in the product one chooses the argument of the j'th factor by specifying that $|\theta - \theta_j| < \pi$. The Fisher-Hartwig conjecture actually states that

$$D_n(\phi) \sim C\,G(\psi)^n\,n^{-\sum \beta_j^2} \tag{14}$$

where C is another nonzero constant. This was established in [11] for the case of one discontinuity and $|\mathcal{R}\,\beta| < \frac{1}{2}$, in [2] (and, independently, in [3]) for an arbitrary (finite) number of discontinuities and all $|\mathcal{R}\,\beta_j| < \frac{1}{2}$, in [4] for one discontinuity and $|\mathcal{R}\,\beta| < 1$, and in [1] for an arbitrary number of discontinuities and either $0 < \mathcal{R}\,\beta_j < 1$ for all j or $-1 < \mathcal{R}\,\beta_j < 0$ for all j. These results have eigenvalue distribution consequences. For example the third one quoted implies that if there is only one discontinuity, and if $|i(\phi - \lambda)| < 1$ (with the obvious meaning of index) for all $\lambda \notin$ range ϕ (where "range ϕ" now means the closure of the range of ϕ) then we have canonical eigenvalue distribution; and furthermore (12) holds.

Without a small index condition results are much more difficult to establish. They require a fine knowledge of the asymptotics of the entries of $T_n(\phi_0)^{-1}$ when ϕ_0 has discontinuities and satisfies a small index condition. The only such results to date establish (14), and so canonical eigenvalue distribution and (12), when there is a single discontinuity–first for $|\mathcal{R}\,\beta| < \frac{5}{2}$ [8], and then for general β [9]. Extending this to more than one discontinuity would require great perseverence, a new idea, or (more likely) both.

References

[1] E. Basor. The extended Fisher-Hartwig conjecture and symbols with multiple jump discontinuities. *To appear.*

[2] E. Basor. A localization theorem for Toeplitz determinants. *Indiana Univ. Math. J.*, 28:975–983, 1979.

[3] A. Böttcher. Toeplitz determinants with piecewise continuous generating function. *Z. Anal. Anw.*, 1:23–39, 1982.

[4] A. Böttcher and B. Silbermann. Toeplitz operators and determinants generated by symbols with one Fisher-Hartwig singularity. *Math. Nachr.*, 127:95–124, 1986.

[5] K. M. Day. Measures associated with Toeplitz matrices generated by the Laurent expansion of rational functions. *Trans. Amer. Math. Soc.*, 209:175–183, 1975.

[6] M. E. Fisher and R. E. Hartwig. Toeplitz determinants: some applications, theorems, and conjectures. *Adv. Chem. Phys.*, 15:333–353, 1968.

[7] I. I. Hirschman, Jr. The spectra of certain Toeplitz matrices. *Illinois J. Math*, 11:145–159, 1967.

[8] R. Libby. Asymptotics of determinants and eigenvalues for Toeplitz matrices associated with certain discontinuous symbols. *Ph.D. Thesis, Univ. of Cal. Santa Cruz*, 1990.

[9] R. Libby. In preparation. 1993.

[10] P. Schmidt and F. Spitzer. The Toeplitz matrices of an arbitrary Laurent polynomial. *Math. Scand.*, 8:15–38, 1960.

[11] H. Widom. Toeplitz determinants with singular generating functions. *Amer. J. Math.*, 95:333–383, 1973.

[12] H. Widom. Eigenvalue distribution of nonselfadjoint Toeplitz matrices and the asymptotics of Toeplitz determinants in the case of nonvanishing index. *Oper. Th.: Adv. and Appl.*, 48:387–421, 1990.

Department of Mathematics
University of California
Santa Cruz, Cal. 95064, U.S.A.

MSC 1991: Primary 47B35, Secondary 15A18,65F15,47A75

Operator Theory:
Advances and Applications, Vol. 71
© 1994 Birkhäuser Verlag Basel/Switzerland

RANDOM HERMITIAN MATRICES AND (NONRANDOM) TOEPLITZ MATRICES[1]

Harold Widom

In this expository article we discuss some questions from the theory of random matrices which have been answered with the help of Toeplitz matrices. These questions concern the statistics of the spacings between eigenvalues of very large hermitian matrices. In certain models of random matrices they lead to the study of the Fredholm determinant of the "sine kernel" $sin(x - y)/\pi(x - y)$ on a finite interval. This kernel, when discretized, becomes the Toeplitz matrix associated with an arc of the circle and Toeplitz methods help to give the desired answers. Much of this is heuristic, and rigorous proofs are yet to be found.

By a "model" of $N \times N$ hermitian matrices we mean, simply, a probability measure on the set of all such matrices. The matrix model we shall be concerned with is the "Gaussian Unitary Ensemble", and is characterized by the following two properties:

(i) The measure is invariant under unitary equivalence.

(ii) The real and imaginary parts of the entries are (as much as they can be, given that the matrix is hermitian) statistically independent.

To be more precise, (i) means that if S is any set of matrices and if U is any unitary matrix then S and $U S U^{-1}$ have the same measure, and (ii) means that if the entries of the matrix are $a_{i,j}, (i,j = 1, \ldots, N)$, then the $a_{i,i}$ and the real and imaginary parts of the $a_{i,j}$ with $i < j$ are all independent random variables. (Loosely speaking, information on any of them gives no information on any of the others.)

It turns out that for $N \geq 3$ there is a two (real) parameter family of such measures. (For the proof of this, and the details for what follows, we refer the reader to Chapters 2 and 5 of [7].) We shall not describe them explicitly since we are interested, not in the entries of the matrix, but rather its eigenvalues. Again loosely speaking, the probability density that the eigenvalues $\lambda_1, \ldots, \lambda_N$ lie in the intervals

$$(x_1, x_1 + d\,x_1), \ldots, (x_N, x_N + d\,x_N)$$

is given by the formula

$$p_N(x_1, \ldots, x_N) = e^{-a \sum x_i^2 + b \sum x_i + c} \prod_{i<j}(x_i - x_j)^2.$$

---

[1] Toeplitz Lecture presented at Tel Aviv University, March, 1993

Here $a > 0$ and $b \in \mathbf{R}$ are two arbitrary parameters and c is chosen so that p_N has total integral 1. The above means, precisely, that if f is any symmetric function of N real variables then the expected value of $f(\lambda_1, \dots, \lambda_N)$ equals

$$\int_{\mathbf{R}^N} f(x_1, \dots, x_N)\, p_N(x_1, \dots, x_N)\, dx_1 \dots dx_N. \tag{1}$$

It is usual to take $b = 0$ (which is equivalent to the trace having expected value 0) and $a = 1$, and we shall do the same.

Let J be a subset of $\mathbf{R}$ and n a nonnegative integer. We denote (again conforming with the notation in [7]) by $E_N(n; J)$ the probability that precisely n of the eigenvalues lie in J. Clearly this quantity is related to the statistics of the distribution of the eigenvalues. We shall discuss $E_N(0; J)$ first.

It follows at once from (1) that

$$E_N(0; J) = c_1 \int_{J^c} \dots \int_{J^c} e^{-\sum x_i^2} \prod_{i<j}(x_i - x_j)^2\, dx_1 \dots dx_N$$

for some constant c_1 (depending only on N), where J^c denotes the complement of J. Now everyone knows that a determinant in which every entry is a sum is equal to a multiple sum of determinants. Analogously a determinant in which every entry is an integral equals a multiple integral whose integrand is a determinant. The integrand in the last multiple integral involves the square of a Vandermonde determinant rather than the determinant. Nevertheless, using this idea we can show (an exercise left for the reader) that the above equals

$$c_2 \det \left(\int_{J^c} e^{-x^2} x^{i+j}\, dx \right)_{i,j\,=\,0,\dots,N-1}$$

where c_2 is another constant. The i, j entry of this determinant is equal to the inner product

$$(x^i e^{-x^2/2}, x^j e^{-x^2/2})_{J^c}, \tag{2}$$

where the subscript indicates that the inner product refers to the space $L_2(J^c)$. This suggests the introduction of the parabolic cylinder functions $\varphi_i(x)$, which are obtained by orthonormalizing the sequence $\{x^i e^{-x^2/2}\}$ over $\mathbf{R}$. (These equal $e^{-x^2/2}$ times the Hermite polynomials $H_i(x)$ times a normalizing constant.) Replacing the entries in the inner product by these functions just amounts to certain row and column operations and so

$$E_N(0; J) = c_3 \det\big((\varphi_i, \varphi_j)_{J^c} \big)_{i,j=0,\dots,N-1} = c_3 \det(\delta_{i,j} - (\varphi_i, \varphi_j)_J)_{i,j=0,\dots,N-1}$$

where c_3 is another constant depending only on N. But if $J = \emptyset$ then clearly $E_N(0, J) = 1$ and, equally clearly, the last determinant equals 1. Hence $c_3 = 1$. Finally, if one has a finite rank operator then the eigenvalues of the operator are equal to the eigenvalues of the associated Gram matrix. It follows that if we define

$$K_N(x, y) := \sum_{i=0}^{N-1} \varphi_i(x)\,\varphi_i(y)$$

then

$$E_N(0, J) = \det\,(I - K_N) \tag{3}$$

where K_N denotes the integral operator on $L_2(J)$ with kernel $K_N(x,y)$ and "det" denotes Fredholm determinant. (Of course the operator has only N nonzero eigenvalues, which are the eigenvalues of the associated Gram matrix $(\varphi_i, \varphi_j)_{i,j=0,\ldots,N-1}$.)

Now we want to think of N as very large. In other words we want to pass to the limit $N \to \infty$. In order to see how to do this without getting a trivial result we look at the "density of eigenvalues". This density, at x, is defined to be the limit of the expected number of eigenvalues in an interval around x divided by the length of the interval, as the length tends to zero. It turns out that the density, $\rho_N(x)$, is exactly $K_N(x,x)$, and its behavior as $N \to \infty$ follows from the known asymptotics of the Hermite polynomials. The result can be stated as

$$\lim_{N \to \infty} \sqrt{\frac{2}{N}} \rho(\sqrt{2N}x) = \begin{cases} \frac{2}{\pi}\sqrt{1-x^2} & \text{if } |x| < 1 \\ 0 & \text{if } |x| > 1, \end{cases} \tag{4}$$

which holds uniformly on compact subsets of $|x| < 1$ and of $|x| > 1$. This is known as the "Wigner semi-circle" law (although it is more a semi-ellipse than a semi-circle). It follows from this in particular that for any J of finite measure $E_N(0; J) \to 0$ as $N \to \infty$, and so to get a nonzero limit the set J must shrink (in fact its size must be of the order $N^{-1/2}$) as $N \to \infty$. (This is known as taking a "scaling limit".) What we do, in fact, is take a fixed bounded set J and consider the quantity $E_N(0; J/\sqrt{2N})$. By (3) this is the Fredholm determinant for the kernel $K_N(x,y)$ acting on $J/\sqrt{2N}$ or, equivalently, the kernel

$$\frac{1}{\sqrt{2N}} K_N(\frac{x}{\sqrt{2N}}, \frac{y}{\sqrt{2N}})$$

on J. But from the asymptotics of the Hermite polynomials, again, one can show

$$\lim_{N \to \infty} \frac{1}{\sqrt{2N}} K_N(\frac{x}{\sqrt{2N}}, \frac{y}{\sqrt{2N}}) = \frac{\sin(x-y)}{\pi(x-y)}.$$

Hence

$$E(0; J) := \lim_{N \to \infty} E_N(0; J/\sqrt{2N}) = \det(I - K)$$

where K is the integral operator with kernel

$$K(x,y) = \frac{\sin(x-y)}{\pi(x-y)} \tag{5}$$

on J. Thus this "sine kernel" is the scaling limit of the kernels $K_N(x,y)$.

More generally, one can show that

$$E_N(n; J) = \frac{(-1)^n}{n!} \frac{d^n}{d\lambda^n} \det(I - \lambda K_N)\big|_{\lambda=1}$$

from which it follows that

$$E(n; J) := \lim_{N \to \infty} E_N(n; J/\sqrt{2N}) = \frac{(-1)^n}{n!} \frac{d^n}{d\lambda^n} \det(I - \lambda K_N)\big|_{\lambda=1} \tag{6}$$

Physicists interpret this quantity $E(n; J)$ as the probability that a very large hermitian matrix, with suitably normalized mean spacing of its eigenvalues, has precisely n eigenvalues in J.

Note that because of the translation-invariance of the sine kernel the quantities $E(n; J)$ are invariant under translation of J. In particular, if J is an interval then it depends only on the length of J. In this case we denote its length by s, think of J as the interval $(0, s)$, and write $E(n; s)$ instead of $E(n; J)$.

Clearly $E(0; s)$ is a positive function of s which decreases from 1 to 0. It is easy to obtain precise information on its behavior near $s = 0$ because the (trace) norm of K tends to 0 as $s \to 0$,

$$\det (I - K) = \exp\{\operatorname{tr} \log(I - K)\},$$

and the power series for $\log(I - K)$ is an asymptotic expansion, as $s \to 0$, in the trace norm. The interesting problem is the behavior of $E(0; s)$ as $s \to \infty$.

Some twenty years ago F. J. Dyson asked the writer of these words whether he knew the asymptotics as $n \to \infty$ of the Toeplitz determinant D_n whose symbol is the characteristic function of an arc of the unit circle. Now Toeplitz determinants with symbols supported on the full circle grow or decrease at most exponentially as $n \to \infty$, but if the support is only part of the circle they decrease even more rapidly. In fact this author was able to answer Dyson's question [10]: If E_α denotes the arc $\alpha \leq \theta \leq 2\pi - \theta$ and ψ_α its characteristic function then for the determinant of order $n + 1$ we have

$$D_n(\psi_\alpha) \sim 2^{1/2} e^{3\zeta'(-1)} (n \sin \frac{\alpha}{2})^{1/4} (\cos \frac{\alpha}{2})^{(n+1)^2}. \tag{7}$$

This used the relationship between orthogonal polynomials on the circle and the unit interval ([8], Theorem 11.5) and a "strong Szegő limit theorem" for ultraspherical polynomials due to Hirschman [5].

Now what Dyson was really interested in was the asymptotics of $E(0; s)$ as $s \to \infty$ and his motive for asking the question was that if one discretizes the kernel $\delta(x - y) - K(x, y)$ then one obtains the Toeplitz matrix $T_n(\psi_\alpha)$. For the i, j entry of this matrix equals $\delta_{i,j}$ minus

$$\frac{\sin \alpha(i - j)}{\pi(i - j)} \tag{8}$$

and if we discretize (5) over $(0, s)$ by setting

$$x = \frac{i\,s}{n}, \quad y = \frac{j\,s}{n} \quad (i, j = 1, \ldots, n)$$

then we obtain the matrix with i, j entry given by (8) with $\alpha = s/n$. It is not unreasonable to suppose, then, that the asymptotics of $\det (I - K)$ (where K acts on $(0, s)$) as $s \to \infty$ can be obtained by replacing α by s/n in (7) and then letting $n \to \infty$. What one gets upon doing this is the formula [2]

$$E(0; s) = \det (I - K) \sim 2^{1/3} e^{3\zeta'(-1)} s^{-1/4} e^{-s^2/8} \quad (s \to \infty). \tag{9}$$

This is accepted as true by physicists but has not been proved to this day. (A very little of it has—see below.) But supposing it is true, what can be said of $E(n; s)$? It is clear from (6) that

$$\frac{E(1; s)}{E(0; s)} = \operatorname{tr} K(I - K)^{-1} \tag{10}$$

and so for $n = 1$ it is a question of determining the asymptotics of the right side. We describe the "physicist's method" used in [1] to do this.

Discretizing replaces the resolvent operator

$$K(I - K)^{-1} = (I - K)^{-1} - I$$

by the matrix $T_n(\psi_\alpha)^{-1} - I$. Now a theorem of Gohberg and Semencul ([4], Theorem III.6.1) expresses the inverse of a Toeplitz matrix $T_n(\psi)$ in terms of the vectors $T_n(\psi) e_0$ and $T_n(\tilde{\psi}) e_0$, where $\tilde{\psi}(z) = \psi(z^{-1})$ and $e_0 = (1, 0, \ldots, 0)$. If $\psi \geq 0$ these vectors are the coefficients (and their complex conjugates) of the suitably normalized Szegö polynomials $P_n(z)$ associated with the weight function ψ. One deduces that if the highest coefficient of $P_n(z)$ equals

$$[D_{n-1}(\psi)/D_n(\psi)]^{1/2}$$

then

$$\mathrm{tr}\, T_n(\psi)^{-1} = \frac{1}{2\pi} \int [2\, z\, P_n'(z)\, \overline{P_n(z)} - (n-1)\, |P_n(z)|^2]\, d\theta.$$

Using the known asymptotics of the orthogonal polynomials one finds that the main contribution to the integral comes from the neighborhood of $z = 1$ and that in fact

$$\mathrm{tr}\, T_n(\psi_\alpha)^{-1} \sim \frac{c^{2n}}{4\sqrt{\pi n(c-1)}}(1 + \varepsilon_\alpha) \qquad (n \to \infty)$$

where $c = \tan\frac{\alpha}{2} + \sec\frac{\alpha}{2}$ and ε_α is a constant which tends to zero as $\alpha \to 0$. Then, if scaling is justified, one obtains for the operator on $(0, s)$

$$\mathrm{tr}\, K(I - K)^{-1} \sim \frac{e^s}{2\sqrt{2\pi s}} \qquad (s \to \infty). \tag{11}$$

This, combined with (9) and (10), gives the asymptotics of $E(1; s)$. (For $E(n; s)$ with $n > 1$, see below.)

The qustion is, what can be proved of all this without using any dubious scaling argument? The answer is, quite a bit but not (yet) everything. One can in fact work with the integral operator K directly, for which there is an analogue of the Gohberg-Semencul result mentioned above. One then encounters the problem of the asymptotics of a certain family of Krein functions–entire function analogues of Szegö polynomials. This can be done [11], and (11) was proved. This approach also yielded the asymptotic result

$$\frac{d}{ds}\log\det(I - K) \sim -\frac{s}{4} \qquad (s \to \infty), \tag{12}$$

from which the first-order asymptotics in (9) follows.

There are other (and, in the long run, surely more important) ways of attacking some of these problems which we now mention. For $\lambda \leq 1$ define the function $\sigma(\lambda; s)$ by

$$\det(I - \lambda K) = \exp\left\{-\int_0^s \sigma(\lambda; t)\, dt\right\}. \tag{13}$$

Then it was shown by Jimbo, Miwa, Môri and Sato [6], where the notation was slightly different, that $\sigma(\lambda; s)$ satisfies a second-order nonlinear differential equation in s, an equation of Painlevé type, which is independent of λ.

The asymptotics of the solutions of this equation have not been rigorously ascertained, but it is easy to show that *if* there is an asymptotic ecpansion of the form

$$\sigma(1;s) \sim \frac{s}{4} + \sum_{k=0}^{\infty} c_k s^{-k} \tag{14}$$

(from (12) we do know that $\sigma(1;s) \sim s/4$) then all the c_k can be determined by simply plugging into the equation. Dyson [2] found this same expansion using difficult (and heuristic) inverse scattering techniques. Notice that integrating (14) extends (9) to a complete asymptotic expansion, except that it cannot give the constant factor. The only way one knows how to determine it is by scaling the Toeplitz formula (7).

We return to $E(n;s)$ for general n. It is immediate from (6) and (13) that

$$\frac{E(n;s)}{E(0;s)} = Q_n\left(\int_0^s \sigma_1(t)\,dt, \ldots, \int_0^s \sigma_n(t)\,dt\right)$$

where Q_n is a polynomial in n variables and

$$\sigma_n(s) := \frac{\partial^n}{\partial\lambda^n}\sigma(\lambda;s)\Big|_{\lambda=1}.$$

Notice that $\sigma_0(s)$ is the same as $\sigma(1;s)$.

We can get differential equations for the various σ_n by differentiating n times with respect to λ the Painlevé equation satisfied by $\sigma(\lambda;s)$, and then setting $\lambda = 1$. What we obtain by doing this is a family of equations

$$L\,\sigma_n(s) = f_n(\sigma_1(s),\ldots,\sigma_{n-1}(s)) \tag{15}$$

where f_n is a polynomial and L is a *linear* second-order differential operator with coefficients depending on the function σ_0 whose asymptotics, given by (14), are known. (All this is heuristic, of course.) One can then successively write down asymptotic expansions for the σ_n and therefore also, eventually, for the $E(n;s)$. However, there are arbitrary constants involved in the solutions of the equations (15). It turns out that for $n > 1$ any particular solution dominates all solutions of the homogeneous equation, so that no furher constants need be determined once one knows σ_1. The σ_1 equation is homogeneous and one of two independent solutions dominates the other, so there is one constant factor (the one multiplying the dominant solution in the representation of σ_1) to be determined. And that constant can be fixed by using the asymptotics of $E(1;s)$ already established by Toeplitz methods. Thus the asymptotics of $E(n;s)$ for any n can in principle be determined.

Finally we mention that there is another route to the asymptotics of $E(n;J)$ or, more precisely, of its ratio with $E(0;J)$, which exploits the known asymptotics as $s \to \infty$ of the individual eigenvalues of the sine kernel on $(0,s)$ [3]. Details can be found in Chapter VII of [9].

References

[1] E. Basor, C. A. Tracy, and H. Widom. Asymptotics of level spacing distributions for random matrices. *Physical Review Letters*, 69:5–8, 1992.

[2] F. J. Dyson. Fredholm determinants and inverse scattering problems. *Comm. Math. Phys*, 47:171–183, 1976.

[3] W. H. J. Fuchs. On the eigenvalues of an integral equation arising in the theory of band-limited signals. *J. Math. Anal. and Applic.*, 9:317–330, 1964.

[4] I. C. Gohberg and I. A. Fel'dman. *Convolution Equations and Projection Methods for their Solution*, volume 41 of *Transl. Math. Monographs*. Amer. Math. Soc., Providence, 1974.

[5] I. I. Hirschman, Jr. The strong Szegö limit theorem for Toeplitz determinants. *Amer. J. Math.*, 88:577–614, 1966.

[6] M. Jimbo, T. Miwa, Y. Môri, and M. Sato. Density of an impenetrable Bose gas and the fifth Painlevé transcendent. *Physica*, 1D:80–158, 1990.

[7] M. L. Mehta. *Random Matrices*. Academic Press, San Diego, 1991.

[8] G. Szegö. *Orthogonal Polynomials*, volume 23 of *Colloq. Publications*. Amer. Math. Soc., New York, 1959.

[9] C. A. Tracy and H. Widom. Introduction to random matrices. In *Proc. 8th Scheveningen Conf.* Springer Lecture Notes in Physics, 1993.

[10] H. Widom. The strong Szegö limit theorem for circular arcs. *Indiana U. Math. J.*, 21:271–283, 1971.

[11] H. Widom. The asymptotics of a continuous analogue of orthogonal polynomials. *J. Approx. Th.*, 1993.

Department of Mathematics
University of California
Santa Cruz, Cal. 95064, U.S.A.

MSC 1991: Primary 15A52,   Secondary 45E10,47B35,47N55

Operator Theory:
Advances and Applications, Vol. 71
© 1994 Birkhäuser Verlag Basel/Switzerland

THE EXTENDED FISHER-HARTWIG CONJECTURE FOR SYMBOLS WITH MULTIPLE JUMP DISCONTINUITIES

Estelle L. Basor and Kent E. Morrison

Dedicated to Harold Widom on his sixtieth birthday

The asymptotic expansions of Toeplitz determinants of certain symbols with multiple jump discontinuities are shown to satisfy a revised version of the conjecture of Fisher and Hartwig.

§1. Introduction

The Toeplitz matrix $T_n[\phi]$ is said to be generated by the function ϕ if

$$T_n[\phi] = (\phi_{i-j}), \quad i,j = 0,\ldots,n-1.$$

where

$$\phi_n = \frac{1}{2\pi} \int_0^{2\pi} \phi(\theta)e^{-in\theta}\,d\theta$$

is the nth Fourier coefficient of ϕ. Define the determinant

$$D_n[\phi] = \det(T_n[\phi]), \quad i,j = 0,\ldots,n-1. \tag{1}$$

The Fisher-Hartwig Conjecture [8] concerns the asymptotic behavior of the determinants of Toeplitz matrices for a certain class of singular symbols. These symbols are of the form

$$\phi(\theta) = b(\theta) \prod_{r=1}^{R} t_{\beta_r}(\theta - \theta_r)u_{\alpha_r}(\theta - \theta_r) \tag{2}$$

where

$$t_\beta(\theta) = \exp[-i\beta(\pi - \theta)], \quad 0 < \theta < 2\pi \tag{3}$$

$$u_\alpha(\theta) = (2 - 2\cos\theta)^\alpha, \quad \operatorname{Re}\alpha > -\frac{1}{2} \tag{4}$$

and $b : \mathbf{T} \to \mathbf{C}$ is a smooth non-vanishing function with zero index. Note that the function ϕ may have jump discontinuities, zeros, and/or singularities.

The first general results on the conjecture, which we will describe shortly, were obtained by Widom [9], who showed that it was true for $\operatorname{Re}\alpha_r > -1/2$ and $\beta_r = 0$ for all r. The conjecture was then extended by several authors to restricted values of the parameters β_r and α_r. In particular, it is true if $|\operatorname{Re}\beta_r| < 1/2$ and $|\operatorname{Re}\alpha_r| < 1/2$, and true in the case of one singularity for arbitrary β where $\alpha = 0$. A history of this work can be found in [7].

Recently, in an investigation of the distribution of eigenvalues of Toeplitz matrices it was shown that in some simple and unexpected cases the original conjecture was false [3, 2]. Earlier Böttcher and Silbermann [5] had also shown that the conjecture did not hold in the case of integer parameter values. A revised conjecture subsuming both kinds of counter-examples was formulated in [3].

The purpose of this paper is to prove the original conjecture in some additional cases, give an example underpinning the revised conjecture, and to discuss the implications for Toeplitz eigenvalues. We begin with a description of the new conjecture which takes into account the possibility of multiple representations of the symbol in the form specified by (2).

Conjecture *Suppose*

$$\phi(\theta) = b^i(\theta) \prod_{r=1}^{R} t_{\beta_r^i}(\theta - \theta_r) u_{\alpha_r^i}(\theta - \theta_r) \tag{5}$$

for values $\beta_1^i, \ldots, \beta_R^i, \alpha_1^i, \ldots, \alpha_R^i$ and smooth nonzero functions $b^i(\theta)$ each with winding number zero for $i = 1, 2, \ldots$. (When $R > 1$ there is a countable number of different representations. Notice that $|b^i|$ is independent of i.) Define

$$\Omega(i) = \sum_{r=1}^{R} \left((\alpha_r^i)^2 - (\beta_r^i)^2 \right) \tag{6}$$

$$\Omega = \max_i \operatorname{Re}[\Omega(i)] \tag{7}$$

$$S = \{i \mid \operatorname{Re}[\Omega(i)] = \Omega\}. \tag{8}$$

Then as $n \to \infty$,

$$D_n[\phi] = \sum_{i \in S} G[b^i]^n n^{\Omega(i)} E[b^i, \alpha_r^i, \beta_r^i, \theta_r] + o(G[|b|]^n n^\Omega) \tag{9}$$

where $G[b^i] = \exp(1/2\pi \int_0^{2\pi} \log b^i(\theta) d\theta)$ is the geometric mean of b^i, $G[|b|]$ is the geometric mean of any of the $|b^i|$, since they are all the same, and where $E[b^i, \alpha_r^i, \beta_r^i, \theta_r]$ is a constant described as follows. Suppressing the superscript i, we factor

$$b(\theta) = G[b] b_+(\exp(i\theta)) b_-(\exp(-i\theta))$$

where b_+ extends analytically inside the unit circle and b_- extends analytically outside the unit circle, and $b_+(0) = b_-(\infty) = 1$. Define

$$
\begin{aligned}
E[b, \alpha_r, \beta_r, \theta_r] \;=\; & E[b] \prod_{r=1}^{R} b_-(\exp(i\theta_r))^{-\alpha_r - \beta_r} b_+(\exp(-i\theta_r))^{-\alpha_r + \beta_r} \\
& \times \prod_{1 \le s \neq r \le R} [1 - \exp(i(\theta_s - \theta_r))]^{-(\alpha_r + \beta_r)(\alpha_s - \beta_s)} \\
& \times \prod_{r=1}^{R} G(1 + \alpha_r + \beta_r) G(1 + \alpha_r - \beta_r) / G(1 + 2\alpha_r), \quad (10)
\end{aligned}
$$

where G is the Barnes G-function, $E[b] = \exp(\sum_{k=1}^{\infty} k s_k s_{-k})$, and $s_k := [\log b(\theta)]_k$.

The original statement of Fisher-Hartwig was very similar except that the function ϕ was assumed to be of one fixed form (2). It is important to note that in the previous work cited only one representation yielded the maximum in Ω. In what follows we will show that (9) holds for a function ϕ with $\alpha = 0$ and $-1 < \operatorname{Re}\beta_r \le 0$ (or with $\alpha = 0$ and $0 \le \operatorname{Re}\beta_r < 1$). This is done in section 2. In section 3 we will show that (9) also holds in some examples and discuss the implications for eigenvalues. The results for section 2 agree with the original conjecture; however, the results of section 3 agree only with the revised conjecture.

§2. Localization theorem for $-1 < \operatorname{Re}\beta_r < 0$

We begin this section by showing that in the case $\alpha_r = 0$ and $-1 < \operatorname{Re}\beta_r \le 0$, there is only one representation in (5). It is thus not surprising that section 2 uses an adaptation of older techniques.

Consider a symbol of the form

$$\phi(\theta) = b(\theta) \prod_{r=1}^{R} t_{\beta_r}(\theta - \theta_r). \quad (11)$$

Other representations of ϕ correspond to changing β_r to $\beta'_r = \beta_r + j_r$, where $j_r \in \mathbf{Z}$ such that $\sum j_r = 0$. Now assume that $-1 < \operatorname{Re} \beta_r < 0$ and we will show that this representation is the only one that gives the maximum in (6). Let $\beta_r = x_r + iy_r$. To maximize (6) is equivalent to minimizing

$$f(j) := \sum_{r=1}^{R-1} (x_r + j_r)^2 + (x_R - \sum_{r=1}^{R-1} j_r)^2 \tag{12}$$

over all $j = (j_1, \ldots, j_{R-1}) \in \mathbf{Z}^{R-1}$. Expanding this out we obtain

$$f(j) = 2 \sum_{r=1}^{R-1} j_r^2 + 2 \sum_{r=1}^{R-1} (x_r - x_R) j_r + 2 \sum_{l<r} j_l j_r. \tag{13}$$

Routine algebra rewrites this as $f(j) = j^T A j + 2 b^T j$, where

$$A = \begin{bmatrix} 2 & 1 & 1 & \ldots & 1 \\ 1 & 2 & 1 & \ldots & 1 \\ & & \vdots & & \\ 1 & 1 & \ldots & 1 & 2 \end{bmatrix} \tag{14}$$

and $b = (x_1 - x_R, \ldots, x_r - x_R, \ldots, x_{R-1} - x_R)$. The matrix A is positive definite, so that we have a standard linear algebra problem of minimizing a function which is quadratic plus linear. The minimum over $\mathbf{R}^{R-1}$ is given by $-A^{-1}b$, but we need the minimum on the integral lattice points. Because of the convexity of the function being minimized, the minimal integer points occur at the those integer points surrounding the minimum $-A^{-1}b$. Computing A^{-1}, (either by noting that A is a circulant matrix, by Cramer's Rule, or by guessing), we find

$$A^{-1} = \frac{1}{R} \begin{bmatrix} R-1 & -1 & -1 & \ldots & -1 \\ -1 & R-1 & -1 & \ldots & -1 \\ & & \vdots & & \\ -1 & -1 & \ldots & -1 & R-1 \end{bmatrix} = I + \frac{1}{R} \begin{bmatrix} -1 & \ldots & -1 \\ & \vdots & \\ -1 & \ldots & -1 \end{bmatrix} \tag{15}$$

Let the minimum in $\mathbf{R}^{R-1}$ be the vector $z = -A^{-1}b$. Then

$$z_r = \frac{R-1}{R} x_r - \frac{1}{R} (x_1 + \ldots + \hat{x}_r + \ldots + x_R), \tag{16}$$

where the hat indicates x_r is omitted. From the assumption that $-1 < x_r < 0$ we see that $-1 < z_r < 1$, and that the minimal integer point j must have $j_r = 0, -1, 1$. The expression

$f(j) = j^T A j + 2b^T j$ is 0 for $j = 0$, and we will show that for any other choice of j the expression is positive. First, this is easy to check for $R = 2$. Next, if any $j_r = 0$, then the problem reduces to the case of $R - 1$ variables. This leaves us the situation in which all the j_r are 1 or -1. If all of them have the same sign, then it is also easy to see that $f(j)$ is positive. Therefore, let us assume that m of the j_r's are -1 and p of them are 1, with $m + p = R - 1$. The third term in $f(j)$, formula (13), is

$$2 \sum_{l<r} j_l j_r = 2\left(\binom{p}{2} + \binom{m}{2} - pm\right) \tag{17}$$

Using the restrictions on the x_r we show that

$$f(j) > 2\left(m + \binom{p}{2} + \binom{m}{2} - pm\right) \tag{18}$$

Now the right side factors as $(p-m)(p-m-1)$, from which we see that it is not possible for positive integer values of p and m to make this expression negative. Thus, $f(j) > 0$ for all integer points j whose components are 0,1, or -1. This means that 0 is the unique minimum for (6) as j ranges over $\mathbf{Z}^{R-1}$.

We now restrict our attention to piecewise continuous symbols of the form

$$\phi(\theta) = b(\theta) \prod_{r=1}^{R} t_{\beta_r}(\theta - \theta_r) \tag{19}$$

where $b(\theta)$ is non-zero, sufficiently smooth, has winding number zero, and $-1 < \operatorname{Re} \beta_r < 0$. The Toeplitz operator $T[\phi]$ is represented by the matrix (ϕ_{i-j}), $i, j \geq 0$. The important properties of the Toeplitz operator $T[\phi]$ and the corresponding finite matrices are summarized in [7]. We list some of these below and state some simple consequences.

Let ℓ_p^μ, $1 < p < \infty$, be the weighted space of sequences $x = \{x_n\}$ satisfying $\sum_{n=1}^{\infty} |x_n|^p (n+1)^{p\mu} < \infty$. The class M_p^μ is the collection of all functions $a \in L^1$ such that the convolution $a * x \in \ell_p^\mu$ for all sequences x with finite support and

$$\|a\|_{M_p^\mu} := \sup\{\|a * x\|_p^\mu / \|x\|_p^\mu : x \neq 0\} < \infty. \tag{20}$$

It is clear that the Toeplitz operator $T[a] = (a_{i-j})$, $i, j \geq 0$, and the Hankel operator $H[a] = (a_{i+j})$, $i, j \geq 0$, are bounded on ℓ_p^μ if $a \in M_p^\mu$, and it also can be shown that M_p^μ is a Banach algebra under the above norm.

The closure of the trigonometric polynomials in M_p^μ is denoted by C_p^μ and is contained in the set of continuous functions in L^1. Likewise, $PC_{p,\mu}$ is the closure of the piecewise constant functions in M_p^μ. While the exact description of $PC_{p,\mu}$ is not known, it is true that if $-1/p < \mu < 1/q$, then a function of the form (19) is in $PC_{p,\mu}$ if b has finite total variation. Also, define $W_{r,s}^{\alpha,\beta}$ to be all functions in $L^1(\mathbf{T})$ satisfying $\{f_n, n \geq 0\} \in \ell_s^\beta$ and $\{f_n, n \leq 0\} \in \ell_r^\alpha$. It is known that $W_{1,1}^{\alpha,\beta} \cap W_{p,q}^{\gamma,\delta}$ is an algebra when $\gamma > 1/q$, $\delta > 1/p$, $\alpha \geq 0$, $\beta \geq 0$, and $1/p + 1/q = 1$. From now on it will be assumed that b does not vanish, has index zero, and is sufficiently smooth.

The next question to address is the invertibility of the Toeplitz operators $T[\phi]$ and the applicability of the finite section method to these operators. The answers are contained in the following theorems.

Theorem 1 *Let $p > 1$, $1/p + 1/q = 1$, $-1/p < \mu < 1/q$. Suppose*

(a) b does not vanish, has index zero and has finite total variation.

(b) $-1/p < \mu + \operatorname{Re} \beta_r < 1/q$ for all r.

Then $T[\phi]$ is invertible on ℓ_p^μ.

Proof. The operators $T[b]$, $T[t_{b_r}]$ are invertible on ℓ_p^μ due to Proposition 6.24 of [7], and thus, by Propositions 6.29 and 6.32 of [7], the operator $T[\phi]$ is Fredholm of index zero, and hence, invertible.

Theorem 2 *Let $p > 1$. Suppose b satisfies (a) of the hypothesis of Theorem 1 and, in addition, $-\operatorname{Re} \beta_r < 1/q$, $-\operatorname{Re} \beta_r - 1/p < \mu < 1/q$. Then $T_n[\phi]$ is invertible for all n sufficiently large and $T_n[\phi]^{-1}$ converges strongly to $T[\phi]^{-1}$ on ℓ_p^μ.*

Proof. This is immediate from Theorem 7.45 in [7].

Theorem 3 *If $p > 2$, $\epsilon > 0$, then the embeddings $\ell_p^\mu \to \ell_2^{\mu-1/2+1/p-\epsilon}$, $\ell_2^\mu \to \ell_q^{\mu-1/2+1/p-\epsilon}$, $\ell_2^\mu \to \ell_p^\mu$, $\ell_q^\mu \to \ell_2^\mu$ are continuous.*

Proof. See [6].

The Hankel operator $H[\phi]$ is represented by the matrix (ϕ_{i+j+1}), $i, j \geq 0$, and enjoys the following properties.

Theorem 4 *Let $p > 2$, $-1/p < \mu < 1/q$. Suppose ϕ, ψ are of the form (19), have no common discontinuities and that there exists a smooth partition of unity f, g such that ϕf and ψg have Fourier coefficients a_n satisfying $\sum_{-\infty}^{\infty} |a_n| n^3 < \infty$. Let $\tilde{\psi}(\theta) = \psi(-\theta)$. Then $H[\phi]H[\tilde{\psi}]$ can be realized as a sum of bounded operators which are compositions of the form*

$$AB : \ell_2^{\mu-1/2+1/p-\epsilon} \to \ell_p^\mu \tag{21}$$

where $B : \ell_2^{\mu-1/2+1/p-\epsilon} \to \ell_2^\mu$, B is trace class, and $A : \ell_2^\mu \to \ell_p^\mu$ for some sufficiently small ϵ.

Proof. Consider the identity found in [1]

$$H[\phi]H[\tilde{\psi}] = H[\phi f]H[\tilde{\psi}] + H[\phi]H[\tilde{g\psi}] + H[\phi]H[\tilde{f}]T[\phi] + H[\phi]H[\tilde{g}]T[\psi]. \tag{22}$$

We will show that the operator $H[h]$ is trace class from $\ell_2^{\mu-1/2+1/p-\epsilon}$ to ℓ_2^μ, where h is one of the functions ϕf, $\tilde{f}$, $\tilde{g}$, or $\tilde{g\psi}$. The operators $H[h]$ can be factored as CD where

$$D : \ell_2^{\mu-1/2+1/p-\epsilon} \to \ell_2^\mu, \ D_{jk} = h_{j+k+1}(j+1) \tag{23}$$

$$C : \ell_2^\mu \to \ell_2^\mu, \ C_{jk} = \frac{\delta_{jk}}{j+1} \tag{24}$$

are both Hilbert-Schmidt, as is easily seen by noting that

$$\sum_{j,k} |a_{j+k+1}|^2 (j+1)^{2+2\mu}(k+1)^{-2\mu+1-2/p+2\epsilon} < \infty. \tag{25}$$

Choosing ϵ sufficiently small, the operators $H[\tilde{\psi}]$, $T[\psi]$, and $T[\phi]$ are bounded on $\ell_2^{\mu-1/2+1/p-\epsilon}$. Finally, the operators $H[\phi f]$ and $H[\phi]$ are bounded from ℓ_2^μ to ℓ_p^μ since the inclusion is bounded from ℓ_2^μ to ℓ_p^μ and the Hankel operators are bounded on ℓ_p^μ.

We now show how the conjecture holds in the case $-1 < \operatorname{Re}\beta_r < 0$ and b sufficiently nice. The proof follows exactly that found in [1] or [4] with the appropriate modifications of the spaces. The idea of using cleverly chosen spaces goes back to Böttcher and Silbermann in [6]. The method used in the previous papers hinges on the fundamental identity

$$T_n[\phi\psi] = T_n[\phi]T_n[\psi] + P_n T[\phi]Q_n T[\psi]P_n + P_n H[\phi]H[\tilde{\psi}]P_n. \tag{26}$$

Here P_n is the projection defined on any ℓ_p^μ as

$$P_n(a_0, a_1, \ldots, a_{n-1}, a_n, \ldots) = (a_0, a_1, \ldots, a_{n-1}, 0, 0, \ldots)$$

and $Q_n = I - P_n$. Formula (26) above can be rewritten as

$$T_n[\psi]^{-1}T_n[\phi]^{-1}T_n[\phi\psi] = I_n + A_n + B_n \tag{27}$$

$$= (I_n + B_n)(I_n + A_n) - B_n A_n \tag{28}$$

$$A_n = T_n[\psi]^{-1}T_n[\phi]^{-1}P_n H[\phi]H[\tilde\psi]P_n \tag{29}$$

$$B_n = T_n[\psi]^{-1}T_n[\phi]^{-1}P_n T[\phi]Q_n T[\psi]P_n. \tag{30}$$

We now assume $\phi(\theta)$ has the form given in (19) such that $b(\theta)$ does not vanish and has index zero. Suppose $b \in W_{1,1}^{3,3}$ and $-1 < \operatorname{Re}\beta_r < 0$. Then we can first find $p > 2$ and μ so that Theorem 2 holds. Consider first the matrix A_n as an operator on $\ell_2^{\mu-1/2+1/p-\epsilon}$ by seeing it as the composition of the following operators:

$$P_n H[\phi]H[\tilde\psi]P_n \ : \ \ell_2^{\mu-1/2+1/p-\epsilon} \to \ell_p^\mu$$

$$T_n[\psi]^{-1}T_n[\phi]^{-1} \ : \ \ell_p^\mu \to \ell_p^\mu$$

$$i \ : \ \ell_p^\mu \to \ell_2^{\mu-1/2+1/p-\epsilon}$$

From Theorem 4, the operator $H[\phi]H[\tilde\psi]$ is trace class with the appropriate choice of ϵ. Thus, the sequence of operators A_n converges in the trace norm. Since B_n converges to zero strongly, the product $B_n A_n$ will also converge to zero in the trace norm. Thus, just as in [1] or [4]

$$\det T_n[\phi]^{-1}T_n[\psi]^{-1}T_n[\phi\psi] \sim \det(I_n + A_n)\det(I_n + B_n). \tag{31}$$

This last statement yields the localization needed to prove (9) by induction on the number of singularities; since A_n converges in the trace norm to an operator A on $\ell_2^{\mu-1/2+1/p-\epsilon}$ and $\det(I_n + B_n) = \det(I_n + B_n')$, where, again as in [1], B_n' converges to an operator B' which is trace class on $\ell_2^{-(\mu-1/2+1/p-\epsilon)}$. To summarize, we have

Theorem 5 *Suppose* $\phi(\theta) = b(\theta)\prod_{r=1}^{R} t_{\beta_r}(\theta - \theta_r)$ *where*

(i) $b(\theta)$ *does not vanish and has index zero,*

(ii) $-1 < \operatorname{Re}\beta_r \leq 0$,

(iii) $b \in W_{1,1}^{3,3}$.

Then (9) holds.

The above localization shows that the asymptotic formula in (9) holds. The fact that the constants agree follows from the validity of the conjecture for $|\operatorname{Re}\beta_r| < 1/2$ and the fact that the constants must be analytic functions of β_r. Also, note that by taking conjugates the conjecture also holds if $0 < \operatorname{Re}\beta_r < 1$.

§3. Some Special Cases

Let us consider $\phi(\theta) = t_{\beta_1}(\theta)t_{\beta_2}(\theta + \pi)$ with any arbitrary β_1 and β_2. A simple calculation shows there is more than one contributing representation in (9) if and only if $\operatorname{Re}\beta_2 - \operatorname{Re}\beta_1$ equals some odd integer. To see this, first note that if we pick one representation in (5), say using β_1 and β_2, then any other must be of the form $\beta_1 + k$ and $\beta_2 - k$ where k is some integer. Let $\beta_1 = u_1 + iv_1$ and $\beta_2 = u_2 + iv_2$. Then

$$\operatorname{Re}\Omega(k) = -2k^2 - 2u_1 k + 2u_2 k - u_1{}^2 + v_1{}^2 - u_2{}^2 + v_2{}^2.$$

This quadratic function of k has exactly two maxima for integer values of k when $u_1 - u_2$ is some odd integer. Conversely, if $u_1 - u_2 = 2j + 1$, then the pairs $\beta_1 + j$, $\beta_2 - j$ and $\beta_1 + j + 1$, $\beta_2 - j + 1$ yield two maximizing representations. If $\operatorname{Re}\beta_2 - \operatorname{Re}\beta_1 = 2j + 1$ for an integer j, then the two maximizing representations are, up to a sign, given by

$$\phi(\theta) = t_{(\beta_1+j)}(\theta)t_{(\beta_2-j)}(\theta + \pi) \tag{32}$$

and

$$\phi(\theta) = (-1)t_{(\beta_1+j+1)}(\theta)t_{(\beta_2-j-1)}(\theta + \pi) \tag{33}$$

Let us assume that we have picked one of those representations as a starting point, i.e. that the β_r's differ by one. Then for this new choice $\operatorname{Re}\beta_2 - \operatorname{Re}\beta_1 = 1$.

The prediction from the Extended Fisher-Hartwig Conjecture is that

$$\begin{aligned} D_n[\phi] \;\sim\; & n^{-\beta_1{}^2-\beta_2{}^2} 2^{2\beta_1\beta_2} G(1 + \beta_1)G(1 - \beta_1)G(1 + \beta_2)G(1 - \beta_2) \\ & \times \left(1 + (-1)^n (2n)^{2(\operatorname{Im}\beta_2 - \operatorname{Im}\beta_1)} \frac{\Gamma(1 + \beta_1)\Gamma(1 - \beta_2)}{\Gamma(-\beta_1)\Gamma(\beta_2)}\right) \end{aligned} \tag{34}$$

We can also easily compute the Fourier coefficients for this function.

$$\phi_n = \begin{cases} \frac{1}{\pi(\beta_2+\beta_1-n)}(\sin\beta_2\pi + \sin\beta_1\pi) & \text{if } n \text{ is even} \\[2mm] \frac{1}{\pi(\beta_2+\beta_1-n)}(\sin\beta_1\pi - \sin\beta_2\pi) & \text{if } n \text{ is odd} \end{cases} \tag{35}$$

Thus, if β_1 and β_2 satisfy $\beta_2 = \beta_1 + 1$, we have

$$\phi_n = \begin{cases} 0 & \text{if } n \text{ is even} \\[2mm] \frac{2\sin\beta_1\pi}{\pi(\beta_2+\beta_1-n)} & \text{if } n \text{ is odd} \end{cases} \tag{36}$$

Now by rearranging the rows and columnms in an obvious way so that even and odd coefficients occur in blocks one can easily see that $D_n[\phi] = 0$ for n odd. For n even, this same arrangement yields

$$D_n[\phi] = D_{n/2}[t_{\gamma_1}]D_{n/2}[t_{\gamma_2}] \tag{37}$$

where $\gamma_1 = \frac{\beta_1+\beta_2+1}{2}$ and $\gamma_2 = \frac{\beta_1+\beta_2-1}{2}$, from which it follows that $\gamma_1 = \beta_1+1$ and $\gamma_2 = \beta_1$. A routine computation shows that this agrees with (9). Thus we have proved (9) in the case that $\beta_2 = \beta_1 + 2j + 1$ and $\theta_2 = \theta_1 + \pi$.

We now turn to some examples which illustrate the implication of the Extended Fisher-Hartwig Conjecture for the distribution of eigenvalues of Toeplitz matrices. Consider the function

$$\phi(\theta) = \begin{cases} e^{i(3\pi/4-\theta)} & \text{if } 0 < \theta < \pi \\[2mm] e^{i(-3\pi/4-\theta)} & \text{if } \pi < \theta < 2\pi \end{cases} \tag{38}$$

If we look at the equation $\det(T_n[\phi] - \lambda I_n) = 0$, we see that the values of the β_r's in the canonical product for $\phi - \lambda$ will be given by the correct choice of the parameters

$$\begin{aligned} \beta_1(\lambda) &= \frac{1}{2\pi i}\log(\frac{\phi(0^-)-\lambda}{\phi(0^+)-\lambda}) \\[2mm] \beta_2(\lambda) &= \frac{1}{2\pi i}\log(\frac{\phi(\pi^-)-\lambda}{\phi(\pi^+)-\lambda}). \end{aligned} \tag{39}$$

A quick computation of the arguments shows that for all functions of the form $\phi - \lambda$ with λ not in the image, either Theorem 5 applies or that $-1/2 < \operatorname{Re}\beta_r < 1/2$, in which case earlier results can be used. Since the determinant does not vanish asymptotically, this implies that the eigenvalues will cluster around the image. The following figure shows that the numerical approximation of the eigenvalues for $T_{50}[\phi]$ nicely illustrates the results of section 2.

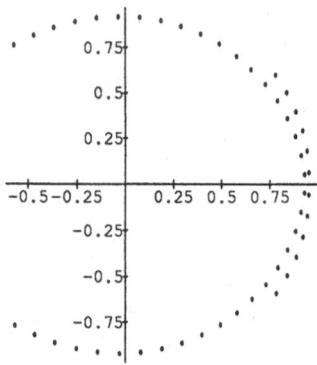

Figure 1. Eigenvalues of $T_{50}[\phi]$.

Another interesting example, described more fully in [2], is one with symbol

$$\phi(\theta) = \begin{cases} \theta + i & \text{if } 0 < \theta < \pi \\ \theta + 2i & \text{if } \pi < \theta < 2\pi \end{cases} \tag{40}$$

This function has two discontinuities and the range is two disjoint line segments. The extended conjecture would imply, as was pointed out earlier, that $\det T_n[\phi - \lambda] \neq 0$, for large n, except when λ is in the image of ϕ or when $-\operatorname{Re}\beta_1(\lambda) + \operatorname{Re}\beta_2(\lambda) = l$, where l is odd. The real parts of $\beta_1(\lambda)$ and $\beta_2(\lambda)$ are given by an appropriate choice of arguments for the logarithms found in formula (39). Let $\lambda = x + iy$ and solve for the arguments. Then take the tangents of both sides of the equation

$$-\operatorname{Re}\beta_1(\lambda) + \operatorname{Re}\beta_2(\lambda) = l$$

to arrive at this cubic equation that x and y must satisfy:

$$2y^3 + 2yx^2 - 4\pi xy - 9y^2 - 3x^2 + 5\pi x + (13 + 2\pi^2)y - 2\pi^2 - 6 = 0. \tag{41}$$

The parameter l disappears here since $\tan 2\pi l = 0$.

The next two pictures show that there are some "stray" eigenvalues that do not lie close to the image.

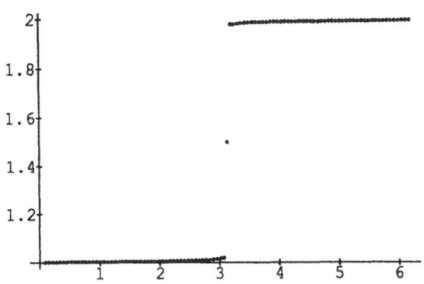

Figure 2. Eigenvalues of $T_{101}[\phi]$.

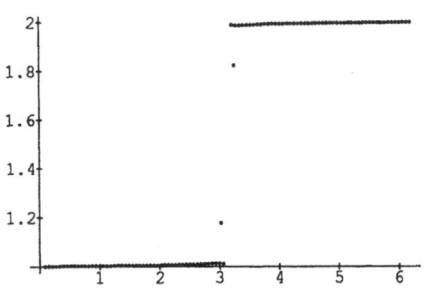

Figure 3. Eigenvalues of $T_{102}[\phi]$.

The next diagram shows a plot of the cubic curve in formula (41). Notice that the stray eigenvalues in the previous plots are near this cubic curve.

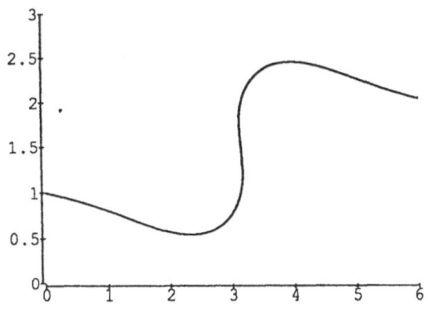

Figure 4. Eigenvalues lie near this cubic curve.

References

[1] E. L. Basor. A localization theorm for Toeplitz determinants. *Indiana Math. J.*, 28:975–983, 1979.

[2] E. L. Basor and K. E. Morrison. The Fisher-Hartwig conjecture and Toeplitz eigenvalues. *Linear Algebra and Its Applications*, 1993.

[3] E. L. Basor and C. A. Tracy. The Fisher-Hartwig conjecture and generalizations. *Phys. A*, 177:167–173, 1991.

[4] E. L. Basor and H. Widom. Toeplitz and Wiener-Hopf determinants with piecewise continuous symbols. *J. Functional Analysis*, 50:387–413, 1983.

[5] A. Böttcher and B. Silbermann. The asymptotic behavior of Toeplitz determinants for generating functions with zeros of integral orders. *Math. Nachr.*, 102:79–105, 1981.

[6] A. Böttcher and B. Silbermann. Toeplitz operators and determinants generated by symbols with one Fisher-Hartwig singularity. *Math. Nachr.*, 127:95–124, 1986.

[7] A. Böttcher and B. Silbermann. *Analysis of Toeplitz Operators*. Springer-Verlag, Berlin, 1990.

[8] M. E. Fisher and R. E. Hartwig. Toeplitz determinants, some applications, theorems and conjectures. *Adv. Chem. Phys.*, 15:333–353, 1968.

[9] H. Widom. Toeplitz determinants with singular generating functions. *Amer. J. Math.*, 95:333–383, 1973.

Department of Mathematics
California Polytechnic State University
San Luis Obispo, CA 93407

E-mail: ebasor@oboe.calpoly.edu and kmorriso@oboe.calpoly.edu

1991 Mathematics Subject Classification. Primary 47B35; Secondary 15A60.

Operator Theory:
Advances and Applications, Vol. 71
© 1994 Birkhäuser Verlag Basel/Switzerland

A RELATIVE TOEPLITZ–HAUSDORFF THEOREM

Hari Bercovici, Ciprian Foias and Allen Tannenbaum

Dedicated to Harold Widom on the occasion of his sixtieth birthday.

We prove an infinite dimensional variant of the Toeplitz-Hausdorff theorem which we found to be useful in the study of structured singular values. We think however that the result is of independent interest, and may find other applications.

Let $\mathcal{H}$ be a complex Hilbert space, and T a bounded linear operator on $\mathcal{H}$. The famous Toeplitz-Hausdorff theorem states that the set

$$W(T) = \{\langle Th, h \rangle : h \in \mathcal{H}, \|h\| = 1\}$$

is a convex subset of the plane; see [1] for an elegant proof. Given an additional operator Q on $\mathcal{H}$, the set $\{\langle Th, h \rangle : h \in \mathcal{H}, \|h\| = 1, Qh = 0\}$ is also seen to be convex by an application of the Toeplitz–Hausdorff theorem in the kernel of Q. The set we are interested in here is a relative numerical range defined as follows:

$$W_Q(T) = \{\lambda = \lim_{n \to \infty} \langle Th_n, h_n \rangle : h_n \in H, \|h_n\| = 1, \lim_{n \to \infty} \|Qh_n\| = 0\}.$$

Observe that there is some similarity between this definition and the characterization given in [2] of the essential numerical range of T. If, for instance, Q has no kernel but is not essentially left invertible then $W_Q(T)$ is contained in the essential numerical range of T.

Our main result is as follows.

* This work was supported in part by grants from the Research Foundation of Indiana University, by the National Science Foundation DMS-8811084, ECS-9001371, ECS-9122106, and by the Air Force Office of Scientific Research AFOSR-90-0024 and AFOSR-90-0053, and by the Army Research Office DAAL03-91-G-0019 and DAAH04-93-G-0332.

THEOREM 1.   *The set $W_Q(T)$ is compact and convex.*

In order to prove this we require a preliminary result, which may be of independent interest when a more precise form of the theorem is needed.

We will denote by $\Re z$ the real part of the complex number z.

LEMMA 1.   *Let Q and T be operators of norm 1 on a Hilbert space $\mathcal{H}$, and let $\alpha, \beta > 0$, $\beta \leq \frac{1}{2}$. Suppose that for every complex number ζ , $|\zeta| = 1$, there exists $h \in \mathcal{H}$ such that*

$$\|h\| = 1 \, , \ \|Qh\| \leq \alpha \, , \ \Re\zeta\langle Th, h\rangle \geq -\beta.$$

Then there exists $h_0 \in \mathcal{H}$ such that

$$\|h_0\| = 1 \, , \ \|Qh_0\| \leq \frac{3\alpha}{\beta} \, , \ \text{and } |\langle Th_0, h_0\rangle| \leq 2\beta.$$

PROOF. We define an auxilliary set by

$$W_\alpha = \{\langle Th, h\rangle : h \in H, \|h\| = 1, \|Qh\| \leq \alpha\},$$

and set $\gamma_\alpha = \inf\{|\lambda| : \lambda \in W_\alpha\}$. Note that $\|Qh\| = \|(Q^*Q)^{1/2}h\|$, $h \in \mathcal{H}$, and so we may assume that $Q \geq 0$. We prove first that W_α is connected. In order to do this, it suffices to show that the set

$$\Omega_\alpha = \{h \in \mathcal{H} : \|h\| = 1, \|Qh\| \leq \alpha\}$$

is connected. Since we must show that any pair of vectors in this set lie in the same component, we may restrict outselves to the case when $\dim \mathcal{H} = 2$. Let then $e_1, e_2 \in \mathcal{H}$ be such that $Qe_j = \lambda_j e_j$, and note that

$$\Omega_\alpha = \{\alpha_1 e_1 + \alpha_2 \alpha_2 : |\alpha_1|^2 + |\alpha_2|^2 = 1, \lambda_1 |\alpha_1|^2 + \lambda_2 |\alpha_2|^2 \leq \alpha\}$$
$$= \{\zeta_1 t_1 e_1 + \zeta_2 t_2 e_2 : |\zeta_1| = |\zeta_2| = 1, t_1 \geq 0, t_2 \geq 0,$$
$$t_1^2 + t_2^2 = 1, \lambda_1 t_1^2 + \lambda_2 t_2^2 \leq \alpha\}.$$

Since the set

$$\{t_1 e_1 + t_2 e_2 : t_1 \geq 0 \, , \ t_2 \geq 0 \, , \ t_1^2 + t_2^2 = 1 \, , \ \lambda_1 t_1^2 + \lambda_1 t_1^2 \leq \alpha\}$$

is obviously connected, we see that W_α is connected as well.

Next fix a positive number $\delta \leq 1$. We claim that for every $\lambda_1, \lambda_2 \in W_\alpha$ such that $|\lambda_1 - \lambda_2| \geq \delta$, the whole segment $[\lambda_1, \lambda_2]$ is contained in the set $W_{\alpha'}$, where $\alpha' = 9\alpha/\delta$. Indeed, fix $h_1, h_2 \in \Omega_\alpha$ such that $\lambda_j = \langle Th_j, h_j\rangle$, $j = 1, 2$. We can write $h_2 = \xi_1 h_1 + \xi_2 h_2'$ with $\|h_2'\| = 1$, $h_2' \perp h_1$, $|\xi_1|^2 + |\xi_2|^2 = 1$, and we claim that

$$|\xi_2| \geq \delta/4.$$

Indeed,

$$\delta \leq |\lambda_1 - \lambda_2|$$
$$= |(1 - |\xi_1|^2)\langle Th_1, h_1 \rangle + \Re\, \xi_1 \bar{\xi_2}[\langle Th_1, h_2' \rangle + \langle Th_2', h_1 \rangle] + |\xi_2|^2 \langle Th_2', h_2' \rangle|$$
$$\leq |\xi_2|^2 |\lambda_1| + 2|\xi_1||\xi_2| + |\xi_2|^2$$
$$\leq 4|\xi_2|.$$

Consequently

$$\|Qh_2'\| \leq \frac{1}{|\xi_2|}(\|Qh_2\| + |\xi_1|\|Qh_1\|) \leq \frac{2\alpha}{|\xi_2|} \leq \frac{8\alpha}{\delta}.$$

The Toeplitz-Hausdorff theorem applied to the space generated by h_1 and h_2 shows that for every $t \in (0, 1)$ there is a unit vector $h' = \eta_1 h_1 + \eta_2 h_2'$ such that $\langle Th', h' \rangle = t\lambda_1 + (1-t)\lambda_2$. Since

$$\|Qh'\| \leq \|Qh_1\| + \|Qh_2'\| \leq \alpha + \frac{8\alpha}{\delta} \leq \frac{9\alpha}{\delta},$$

we conclude that $t\lambda_1 + (1 - t)\lambda_2 \in W_{\alpha'}$.

We proceed now to prove the lemma. We may assume that $\gamma_\alpha > 2\beta$, since otherwise the result is trivial. Observe that the set

$$\left\{ \frac{\lambda}{|\lambda|} : \lambda \in W_\alpha \right\}$$

is an arc of the unit circle. Moreover, the hypothesis of the lemma implies that the length of this arc is at least $\pi - 2\sin^{-1}\beta$. Thus there exist $\lambda_1, \lambda_2 \in W_\alpha$ with $1 \geq |\lambda_1|, |\lambda_2| \geq 2\beta$ and

$$|\arg \lambda_1 - \arg \lambda_2| = \pi - 2\sin^{-1}\beta.$$

Clearly then,

$$|\lambda_1 - \lambda_2| \geq 2\beta \left| \frac{\lambda_1}{|\lambda_1|} \right| = 4\beta\sqrt{1 - \beta^2} \geq 3\beta,$$

and there is $\lambda \in [\lambda_1, \lambda_2]$ with $|\lambda| \leq \beta$.

The observation above, applied to $\delta = 3\beta$, implies that $\lambda \in W_{3\alpha/\beta}$. $\qquad\square$

PROOF OF THEOREM 1. We may assume that $\|Q\| \leq 1$ and $\|T\| < \frac{1}{2}$, so $\|T - \lambda\| \leq 1$ for $|\lambda| \leq \frac{1}{2}$. Let $\lambda \in \mathrm{co}W_Q(T)$ (the convex hull of $W_Q(T)$), and fix $\beta > 0$. For every unimodular complex number ζ there exists a sequence $h_n \in H$, $\|h_n\| = 1$ such that $\|Qh_n\| \to 0$ and $\Re\, \zeta\langle(T - \lambda)h_n, h_n \rangle \geq -\beta$. Then the Lemma implies the existence of vectors h_n', $\|h_n'\| = 1$, $\|Qh_n'\| \to 0$ such that $|\langle(T - \lambda)h_n', h_n' \rangle| \leq 2\beta$. Thus we can find points in $W_Q(T)$ arbitrarily close to λ. To conclude the proof we just observe that $W_Q(T)$ is a closed set. $\qquad\square$

REFERENCES

[1.] P.R. Halmos,*A Hilbert Space Problem Book*, Springer-Verlag, New York, 1982.

[2.] P.A. Fillmore, J.G. Stampfli, and J.P. Williams, "On the essential numerical range, the essential spectrum, and a problem of Halmos," *Acta Sci. Math. (Szeged)* **33**(1972), 179–192.

Hari Bercovici and Ciprian Foias
Department of Mathematics
Indiana University
Bloomington, Indiana 47405

Allen Tannenbaum
Department of Electrical Engineering
University of Minnesota
Minneapolis, Minnesota 55455

MSC 1991: 47A12

Operator Theory:
Advances and Applications, Vol. 71
© 1994 Birkhäuser Verlag Basel/Switzerland

OPERATOR-VALUED SZEGÖ-WIDOM LIMIT THEOREMS

Albrecht Böttcher and Bernd Silbermann

Dedicated to Harold Widom on his sixtieth birthday

We extend the strong Szegö-Widom limit theorem on the asymptotic behavior of large Toeplitz determinants generated by matrix-valued functions to the case of operator-valued generating functions. The approach of this paper relies heavily on the Fredholm theory of Toeplitz and Hankel operators with operator-valued symbols and is based upon new results on the finite section method for Toeplitz operators induced by operator-valued functions.

1. Introduction. Let $\mathcal{H}$ be a separable Hilbert space and let $\mathcal{L}(\mathcal{H})$ denote the Banach algebra of all bounded linear operators on $\mathcal{H}$. Given a sequence $\{\hat{a}_n\}_{n \in \mathbb{Z}}$ of operators $\hat{a}_n \in \mathcal{L}(\mathcal{H})$, let T_N stand for the Toeplitz matrix

$$T_N = \begin{pmatrix} \hat{a}_0 & \hat{a}_{-1} & \dots & \hat{a}_{-N} \\ \hat{a}_1 & \hat{a}_0 & \dots & \hat{a}_{-N+1} \\ \vdots & \vdots & & \vdots \\ \hat{a}_N & \hat{a}_{N-1} & \dots & \hat{a}_0 \end{pmatrix}.$$

The problem of describing the asymptotic behavior of an appropriately defined determinant of T_N as $N \to \infty$ has been extensively studied for a long time and by now at least in the scalar (dim $\mathcal{H} = 1$) and matrix (dim $\mathcal{H} < \infty$) cases resulted in a fairly complete theory (see [5],[11], [2], [3] for an account of this topic).

Much less is known in the operator case (dim $\mathcal{H} = \infty$), and first steps into this direction have been made only recently by Gohberg and Kaashoek [6], who established an operator-valued version of the so-called first Szegö limit theorem. Viz, they computed the limit of $\det T_N/\det T_{N-1}$ as $N \to \infty$ in case the function

$$a(t) = \sum_{n \in \mathbb{Z}} t^n \hat{a}_n \quad (t \in \mathbb{T} := \{z \in \mathbb{C} : |z| = 1\})$$

is sufficiently smooth and takes on positive definite values. Their approach is heavily based on factorization arguments and Schur complement techniques.

Our purpose here is to derive descriptions of the asymptotic behavior of the quotient $\det T_N / \det T_{N-1}$ and of $\det T_N$ alone by means of convergence results for the finite section method for infinite Toeplitz matrices with operator-valued entries. The latter idea has been pursued by several authors for many years, and the approach based upon the finite section method will not only prove to be an appropriate tool for removing the self-adjointness of $a(t)$ and for relaxing the smoothness conditions inquired in [6], but will also turn out to be powerful enough to produce operator-valued versions of the so-called strong Szegö-Widom limit theorem.

2. Toeplitz and Hankel operators with operator-valued symbols.

A function $a : \mathbb{T} \to \mathcal{L}(\mathcal{H})$ is said to belong to $L^\infty(\mathcal{L}(\mathcal{H}))$ if it is weakly measurable and

$$\|a\|_\infty := \operatorname*{ess\,sup}_{t \in \mathbb{T}} \|a(t)\|_{\mathcal{L}(\mathcal{H})} < \infty.$$

The Fourier coefficients of a function $a \in L^\infty(\mathcal{L}(\mathcal{H}))$ are defined by

$$\hat{a}_n := \frac{1}{2\pi} \int_0^{2\pi} e^{-in\theta} a(e^{i\theta}) d\theta \quad (n \in \mathbb{Z}).$$

Denote by $l^2(\mathcal{H})$ the Hilbert space of all sequences $f = \{f_n\}_{n=0}^\infty$ with values $f_n \in \mathcal{H}$ for which

$$\|f\|^2 := \sum_{n=0}^\infty \|f_n\|_{\mathcal{H}}^2 < \infty.$$

Generalizing a classical result by Brown and Halmos, Page [8] showed that if $a \in L^\infty(\mathcal{L}(\mathcal{H}))$, then the Toeplitz operator $T(a)$ acting on $l^2(\mathcal{H})$ by the rule

$$(T(a)f)_j = \sum_{k=0}^\infty \hat{a}_{j-k} f_k \quad (j = 0, 1, 2, \ldots)$$

is bounded and its norm equals $\|a\|_\infty$. The function a is usually referred to as the symbol of the operator $T(a)$.

The Hankel operator $H(a)$ induced by a function a in $L^\infty(\mathcal{L}(\mathcal{H}))$ is the bounded [8] operator on $l^2(\mathcal{H})$ defined by

$$(H(a)f)_j = \sum_{k=0}^\infty \hat{a}_{j+k+1} f_k \quad (j = 0, 1, 2, \ldots).$$

Fredholm theory of Toeplitz operators and compactness properties of Hankel operators are glued together by the two easily verifiable identities

$$T(ab) = T(a)T(b) + H(a)H(\tilde{b}), \tag{1}$$
$$H(ab) = T(a)H(b) + H(a)T(\tilde{b}), \tag{2}$$

where a and b are in $L^\infty(\mathcal{L}(\mathcal{H}))$, and $\tilde{b} \in L^\infty(\mathcal{L}(\mathcal{H}))$ is given by $\tilde{b}(t) = b(1/t)$ for $t \in \mathbb{T}$; note that $(\tilde{b})_n^{\widehat{}} = \hat{b}_{-n}$ for all $n \in \mathbb{Z}$ (see [11] or [3, 2.14]).

As usual, a bounded linear operator is said to be Fredholm if it is invertible modulo compact operators. The well known Hartman-Wintner theorem extends to the operator-valued case: if $a \in L^\infty(\mathcal{L}(\mathcal{H}))$ and $T(a)$ is Fredholm on $l^2(\mathcal{H})$, then a is invertible in

$L^\infty(\mathcal{L}(\mathcal{H}))$; this can be proved by arguments similar to the ones used in the matrix case in [2, proof of Theorem 2.30]. The interesting point in the Fredholm theory of Toeplitz operators is to single out those symbols a invertible in $L^\infty(\mathcal{L}(\mathcal{H}))$ which indeed generate Fredholm Toeplitz operators.

Given a separable Hilbert space $\mathcal{K}$, we denote by $\mathcal{C}_1(\mathcal{K})$, $\mathcal{C}_2(\mathcal{K})$, $\mathcal{C}_\infty(\mathcal{K})$ the trace class, Hilbert-Schmidt, and compact operators on $\mathcal{K}$, respectively. The trace class and the Hilbert-Schmidt norms will be denoted by $\|.\|_1$ and $\|.\|_2$. For properties of operators in $\mathcal{C}_p(\mathcal{K})$ we refer the reader to [3] and the literature cited there. We here fix only one result, which will be repeatedly used in the following.

LEMMA 2.1 *Let $\mathcal{K}$ be a separable Hilbert space and suppose $K \in \mathcal{C}_p(\mathcal{K})$, where p is 1,2, or ∞. If B_n and C_n are bounded linear operators on $\mathcal{K}$ such that $B_n \to B$ and $C_n^* \to C^*$ strongly as $n \to \infty$, then*

$$\|B_n K C_n - BKC\|_p \to 0 \quad as \quad n \to \infty.$$

Proof. See e.g. [11] or [7, Theorem III.6.3]. ∎

One can readily check that a function a in $L^\infty(\mathcal{L}(\mathcal{H}))$ induces a Hankel operator $H(a)$ belonging to $\mathcal{C}_2(l^2(\mathcal{H}))$ if and only if $\hat{a}_n \in \mathcal{C}_2(\mathcal{H})$ for all $n \geq 1$ and

$$\|H(a)\|_2^2 = \sum_{n=1}^\infty n\|\hat{a}_n\|_2^2 < \infty.$$

The characterization of all Hankel operators in $\mathcal{C}_\infty(l^2(\mathcal{H}))$ is a more delicate problem. Let $H^\infty(\mathcal{L}(\mathcal{H}))$ refer to the closed subalgebra of $L^\infty(\mathcal{L}(\mathcal{H}))$ consisting of all functions whose Fourier coefficients with negative indices are the zero operator. Furthermore, let $C(\mathcal{C}_\infty(\mathcal{H}))$ stand for the algebra of all continuous functions of $\mathbb{T}$ into $\mathcal{C}_\infty(\mathcal{H})$. Finally, put

$$C(\mathcal{C}_\infty(\mathcal{H})) + H^\infty(\mathcal{L}(\mathcal{H})) := \{f + g : f \in C(\mathcal{C}_\infty(\mathcal{H})), \ g \in H^\infty(\mathcal{L}(\mathcal{H}))\}.$$

The famous Hartman compactness criterion for Hankel operators was extended by Page [8] to the operator-valued case: if $a \in L^\infty(\mathcal{L}(\mathcal{H}))$, then $H(a)$ belongs to $\mathcal{C}_\infty(l^2(\mathcal{H}))$ if and and only if $\tilde{a} \in C(\mathcal{C}_\infty(\mathcal{H})) + H^\infty(\mathcal{L}(\mathcal{H}))$, where, as above, $\tilde{a}(t) := a(1/t)$ for $t \in \mathbb{T}$. We remark that the Hartman-Page criterion along with the identity (2) implies that both

$$C(\mathcal{C}_\infty(\mathcal{H})) + H^\infty(\mathcal{L}(\mathcal{H})) \quad \text{and} \quad C(\mathcal{C}_\infty(\mathcal{H})) + (H^\infty(\mathcal{L}(\mathcal{H})))^* \tag{3}$$

are closed subalgebras of $L^\infty(\mathcal{L}(\mathcal{H}))$ (in the scalar case the first to draw this conclusion was Sarason).

Another important consequence of the Hartman-Page criterion is the following operator-valued version of a theorem by Douglas, which was proved in [2, p. 90].

THEOREM 2.2 *Let $a \in C(\mathcal{C}_\infty(\mathcal{H})) + H^\infty(\mathcal{L}(\mathcal{H}))$. Then $T(a)$ is Fredholm on $\mathcal{L}(l^2(\mathcal{H}))$ if and only if a is invertible in $C(\mathcal{C}_\infty(\mathcal{H})) + H^\infty(\mathcal{L}(\mathcal{H}))$.*

Proof. If a is invertible in $C(\mathcal{C}_\infty(\mathcal{H})) + H^\infty(\mathcal{L}(\mathcal{H}))$, then, by (1) and the Hartman-Page criterion,

$$T(a)T(a^{-1}) - I = H(a)H(\tilde{a}^{-1}) \in \mathcal{C}_\infty(l^2(\mathcal{H})),$$

$$T(a^{-1})T(a) - I = H(a^{-1})H(\tilde{a}) \in \mathcal{C}_\infty(l^2(\mathcal{H})),$$

which implies that $T(a)$ is Fredholm.

Conversely, suppose now that $T(a)$ is Fredholm. Then a is invertible in $L^\infty(\mathcal{L}(\mathcal{H}))$, by the Hartman-Wintner theorem. We so deduce from (2) that

$$0 = H(\tilde{a}^{-1}\tilde{a}) = T(\tilde{a}^{-1})H(\tilde{a}) + H(\tilde{a}^{-1})T(a),$$

and since $H(\tilde{a})$ is compact owing to the Hartman-Page criterion and $T(\tilde{a}^{-1})$ is bounded, it follows that $H(\tilde{a}^{-1})T(a)$ must also be compact. Let B be any right regularizer of $T(a)$, i.e. $T(a)B = I + K$ with $K \in \mathcal{C}_\infty(l^2(\mathcal{H}))$. Then

$$H(\tilde{a}^{-1}) = H(\tilde{a}^{-1})T(a)B - H(\tilde{a}^{-1})K \in \mathcal{C}_\infty(l^2(\mathcal{H})),$$

whence $a^{-1} \in C(\mathcal{C}_\infty(\mathcal{H})) + H^\infty(\mathcal{C}_\infty(\mathcal{H}))$ due to the Hartman-Page criterion.   ∎

Presently we do not know a nice index formula for Fredholm Toeplitz operators with symbols in $C(\mathcal{C}_\infty(\mathcal{H})) + H^\infty(\mathcal{L}(\mathcal{H}))$ in case $\dim \mathcal{H} = \infty$. For symbols of the form $a = I + b$ with $b \in C(\mathcal{C}_\infty(\mathcal{H}))$ we shall provide an index formula in Chapter 6.

3. Finite section method. Let P_N $(N = 0, 1, 2, \ldots)$ denote the projection on $l^2(\mathcal{H})$ defined by

$$P_N : \{f_0, f_1, f_2, \ldots\} \mapsto \{f_0, \ldots, f_N, 0, 0, \ldots\}.$$

We say that the finite section method is applicable to an invertible operator $A \in \mathcal{L}(l^2(\mathcal{H}))$ if there is an $N_0 \geq 0$ such that the compressions ("finite sections") $P_N A P_N | \text{Im } P_N$ are invertible for all $N \geq N_0$ and

$$\sup_{N \geq N_0} \|(P_N A P_N)^{-1} P_N\|_{\mathcal{L}(l^2(\mathcal{H}))} < \infty.$$

In that case $(P_N A P_N)^{-1} P_N$ converges strongly to A^{-1} as $N \to \infty$.

The following simple proposition is well known as a magic wand and was discovered and employed by many people (see e.g. [4], [1], [9], [2], [3]). It will also play a key role in all what follows here.

PROPOSITION 3.1 *Let X be a linear space, let $A : X \to X$ be a linear and invertible operator, and let P and Q be complementary projections on X (i.e. $P + Q = I$). Then the compression $PAP | \text{Im } P$ is invertible if and only if so is the compression $QA^{-1}Q | \text{Im } Q$; in that case*

$$(PAP)^{-1}P = PA^{-1}P - PA^{-1}Q(QA^{-1}Q)^{-1}QA^{-1}P.$$

Proof. Straightforward (see e.g. [3, 7.15]).   ∎

Let us state a first corollary of the preceding proposition.

PROPOSITION 3.2 *If $a \in L^\infty(\mathcal{L}(\mathcal{H}))$, then $T(\tilde{a})$ is invertible on $l^2(\mathcal{H})$ if and only if a is invertible in $L^\infty(\mathcal{L}(\mathcal{H}))$ and $T(a^{-1})$ is invertible on $l^2(\mathcal{H})$.*

Proof. Let $X = l_{\mathbb{Z}}^2(\mathcal{H})$ denote the Hilbert space of all doubly-infinite sequences $\{f_n\}_{n \in \mathbb{Z}}$ such that $f_n \in \mathcal{H}$ for all n and $\sum_{n \in \mathbb{Z}} \|f_n\|_{\mathcal{H}}^2 < \infty$. For $a \in L^\infty(\mathcal{L}(\mathcal{H}))$, the Laurent operator $A = L(a)$ is defined on X by

$$(L(a)f)_j = \sum_{k \in \mathbb{Z}} \hat{a}_{j-k} f_k \quad (j \in \mathbb{Z}),$$

and we denote by P and Q the following projections on X:

$$P: \{f_n\}_{n \in \mathbb{Z}} \mapsto \{\ldots, 0, 0, f_0, f_1, f_2, \ldots\},$$
$$Q: \{f_n\}_{n \in \mathbb{Z}} \mapsto \{\ldots f_{-2}, f_{-1}, 0, 0, 0, \ldots\}.$$

Then $T(\tilde{a})$ and $T(a^{-1})$ may be identified with $QAQ \,|\, \operatorname{Im} Q$ and $PA^{-1}P \,|\, \operatorname{Im} P$, respectively. The assertion so follows immediately from the operator-valued version of the Hartman-Wintner theorem and Proposition 3.1. ∎

Here is another direct consequence of Proposition 3.1. It tells us in particular that the finite section method is always applicable to inverses of Toeplitz operators.

THEOREM 3.3 *Let $a \in L^\infty(\mathcal{L}(\mathcal{H}))$ and suppose $T(a)$ is invertible on $l^2(\mathcal{H})$. Then the operator $P_N T^{-1}(a) P_N \,|\, \operatorname{Im} P_N$ is invertible for all $N \geq 0$ and the finite section method is applicable to $T^{-1}(a)$.*

Proof. Put $Q_N = I - P_N$. Since $Q_N T(a) Q_N \,|\, \operatorname{Im} Q_N$ is unitarily equivalent to $T(a)$ in an obvious way, Proposition 3.1 implies that $P_N T^{-1}(a) P_N \,|\, \operatorname{Im} P_N$ is invertible for all N and that

$$(P_N T^{-1}(a) P_N)^{-1} P_N = P_N T(a) P_N - P_N T(a) Q_N (Q_N T(a) Q_N)^{-1} Q_N T(a) P_N, \quad (4)$$

whence

$$\sup_{N \geq 0} \|(P_N T^{-1}(a) P_N)^{-1} P_N\| \leq \|a\|_\infty + \|a\|_\infty^2 \|T^{-1}(a)\|. \quad ∎$$

The following theorem is one of our key results.

THEOREM 3.4 *Let $a \in C(\mathcal{C}_\infty(\mathcal{H})) + H^\infty(\mathcal{L}(\mathcal{H}))$. Then the finite section method is applicable to $T(a)$ if and only if both $T(a)$ and $T(\tilde{a})$ are invertible.*

Proof. The "only if" part of the theorem is well known (see e.g. [5, Lemma III.1.1]). For the reader's convenience, we repeat a proof here. Thus, let us show that $T(\tilde{a})$ is invertible if the finite section method is applicable to $T(a)$. Define the operators W_N on $l^2(\mathcal{H})$ by

$$W_N: \{f_0, f_1, f_2, \ldots\} \mapsto \{f_N, \ldots, f_0, 0, 0, \ldots\}.$$

Then $W_N^2 = P_N$, and so

$$W_N T(a) W_N \,|\, \operatorname{Im} P_N = (W_N \,|\, \operatorname{Im} P_N)(P_N T(a) P_N \,|\, \operatorname{Im} P_N)(W_N \,|\, \operatorname{Im} P_N)$$

is invertible for all sufficiently large N, for $N \geq N_0$ say, and

$$M := \sup_{N \geq N_0} \|(W_N T(a) W_N)^{-1} P_N\| < \infty.$$

Hence $\|P_N f\| \leq M \|W_N T(a) W_N f\|$ for all $f \in l^2(\mathcal{H})$. Passing to the limit $N \to \infty$ and taking into account that $W_N T(a) W_N = P_N T(\tilde{a}) P_N$, we obtain that $\|f\| \leq M \|T(\tilde{a}) f\|$ for all $f \in l^2(\mathcal{H})$. In a similar way one can show that $\|f\| \leq M \|T(\tilde{a})^* f\|$ for all $f \in l^2(\mathcal{H})$. This proves that $T(\tilde{a})$ is invertible.

Suppose now that $T(a)$ and $T(\tilde{a})$ are invertible. By virtue of (1) and the Hartman-Page criterion,

$$I - T(a^{-1}) T(a) = H(a^{-1}) H(\tilde{a}) \in \mathcal{C}_\infty(l^2(\mathcal{H}))$$

and thus, $T^{-1}(a) = T(a^{-1}) + K$, where $K = H(a^{-1}) H(\tilde{a}) T^{-1}(a) \in \mathcal{C}_\infty(l^2(\mathcal{H}))$. Put $Q_N = I - P_N$. Then

$$Q_N T^{-1}(a) Q_N = Q_N T(a^{-1}) Q_N + Q_N K Q_N,$$

and since $T(a^{-1})$ is invertible due to Proposition 3.2, $Q_N T(a^{-1}) Q_N \,|\, \mathrm{Im}\, Q_N$ is unitarily equivalent to $T(a^{-1})$, and $\|Q_N K Q_N\| \to 0$ as $N \to \infty$ (Lemma 2.1), it follows that $Q_N T^{-1}(a) Q_N \,|\, \mathrm{Im}\, Q_N$ is invertible for all sufficiently large N, say $N \geq N_0$, and that

$$R := \sup_{N \geq N_0} \|(Q_N T^{-1}(a) Q_N)^{-1} Q_N\| < \infty.$$

From Proposition 3.1 we so infer that $P_N T(a) P_N \,|\, \mathrm{Im}\, P_N$ is invertible for all $N \geq N_0$ and that

$$(P_N T(a) P_N)^{-1} P_N = P_N T^{-1}(a) P_N - P_N T^{-1}(a) Q_N (Q_N T^{-1}(a) Q_N)^{-1} Q_N T^{-1}(a) P_N, \quad (5)$$

which implies that

$$\sup_{N \geq N_0} \|(P_N T(a) P_N)^{-1} P_N\| \leq \|T^{-1}(a)\| + R \|T^{-1}(a)\|^2. \quad \blacksquare$$

The preceding theorem is the theorem we need here. We remark, however, that the finite section method can be studied for more general classes of operator-valued symbols; we shall return to this topic elsewhere.

4. The first Szegö-Widom limit theorem. Statement and proof of the first Szegö-Widom limit theorem require some preliminaries.

Let $a \in L^\infty(\mathcal{L}(\mathcal{H}))$ and suppose $T(a)$ is invertible. From Theorem 3.3 we know that then $P_N T^{-1}(a) P_N \,|\, \mathrm{Im}\, P_N$ is invertible for all $N \geq 0$. Put

$$\Delta_N(a) = (P_N T^{-1}(a) P_N)^{-1} P_N.$$

The operator $\Delta_0(a) \in \mathcal{L}(\mathrm{Im}\, P_0) = \mathcal{L}(\mathcal{H})$ is simply denoted by $\Delta(a)$ and referred to as the right multiplicative diagonal of a (see [6]). We first need some properties of $\Delta_N(a)$ and related operators.

Fix an orthonormal basis $\{e_n\}_{n=0}^{\infty}$ in $\mathcal{H}$ and denote by S_k ($k = 0, 1, 2, \ldots$) the orthogonal projection of $\mathcal{H}$ onto the linear span of $e_0, \ldots, e_k$. Given $b \in L^{\infty}(\mathcal{L}(\mathcal{H}))$, we define $S_k b S_k \in L^{\infty}(\mathcal{L}(\mathcal{H}))$ by

$$(S_k b S_k)(t) = S_k b(t) S_k \quad (t \in \mathbb{T}).$$

The essential range $\mathcal{R}(b)$ of a function $b \in L^{\infty}(\mathcal{L}(\mathcal{H}))$ is defined as the spectrum of b in the Banach algebra $L^{\infty}(\mathcal{L}(\mathcal{H}))$.

LEMMA 4.1 *Suppose* $b \in L^{\infty}(\mathcal{L}(\mathcal{H}))$ *and the essential range* $\mathcal{R}(b)$ *of* b *is a compact subset of* $\mathcal{C}_p(\mathcal{H})$, *where* $p \in \{1, 2, \infty\}$. *Then as* $k \to \infty$,

$$\|b - S_k b S_k\|_p := \operatorname*{ess\ sup}_{t \in \mathbb{T}} \|b(t) - S_k b(t) S_k\|_p \to 0.$$

Proof. Given $\varepsilon > 0$, we can find finitely many $v_1, \ldots, v_n$ in $\mathcal{R}(b)$ such that for each $v \in \mathcal{R}(b)$ there exists a $j \in \{1, \ldots, n\}$ with $\|v - v_j\|_p < \varepsilon$. Clearly,

$$\begin{aligned}
\|v - S_k v S_k\|_p &\leq \|v - v_j\|_p + \|v_j - S_k v_j S_k\|_p + \|S_k\|^2 \|v - v_j\|_p \\
&= \|v - v_j\|_p + \|v_j - S_k v_j S_k\|_p.
\end{aligned}$$

Since $S_k = S_k^* \to I$ strongly on $\mathcal{H}$ and since the operators v_j are compact, we infer from Lemma 2.1 that there is a $k_0 \geq 0$ such that $\|v_j - S_k v_j S_k\|_p < \varepsilon$ for all $j \in \{1, \ldots, n\}$ and all $k \geq k_0$. Hence, if $k \geq k_0$ then $\|v - S_k v S_k\|_p < 3\varepsilon$ for all $v \in \mathcal{R}(b)$. ∎

LEMMA 4.2 *Let* $a = I + b \in L^{\infty}(\mathcal{L}(\mathcal{H}))$ *and let* $T(a)$ *be invertible. Suppose* $\hat{b}_n \in \mathcal{C}_2(\mathcal{H})$ *for all* $n \in \mathbb{Z}$ *and* $\sum_{n \in \mathbb{Z}} \|\hat{b}_n\|_2^2 < \infty$. *Define* $a_k := I + S_k b S_k$ *and let* $N \geq 0$ *be any integer.*

(a) *We have*

$$P_N T(a) - P_N \in \mathcal{C}_2(l^2(\mathcal{H})), \qquad T(a) P_N - P_N \in \mathcal{C}_2(l^2(\mathcal{H})), \tag{6}$$
$$P_N T^{-1}(a) - P_N \in \mathcal{C}_2(l^2(\mathcal{H})), \quad T^{-1}(a) P_N - P_N \in \mathcal{C}_2(l^2(\mathcal{H})); \tag{7}$$

(b) *we also have*

$$\Delta_N(a) - P_N \in \mathcal{C}_2(l^2(\mathcal{H})), \tag{8}$$
$$\Delta(a) - I \in \mathcal{C}_2(\mathcal{H}), \tag{9}$$
$$\Delta(a) - I - \hat{b}_0 \in \mathcal{C}_1(\mathcal{H}); \tag{10}$$

(c) *and if, in addition,* $\mathcal{R}(b)$ *is a compact subset of* $\mathcal{C}_{\infty}(\mathcal{H})$, *then as* $k \to \infty$,

$$\|P_N T^{\pm 1}(a_k) - P_N T^{\pm 1}(a)\|_2 \to 0, \tag{11}$$
$$\|T^{\pm 1}(a_k) P_N - T^{\pm 1}(a) P_N\|_2 \to 0, \tag{12}$$
$$\|\Delta_N(a_k) - \Delta_N(a)\|_2 \to 0. \tag{13}$$

Proof. (a) The matrix representation of $P_N T(a)$ is

$$\begin{pmatrix} I + \hat{b}_0 & \hat{b}_{-1} & \dots & \hat{b}_{-N} & \hat{b}_{-N-1} & \dots \\ \hat{b}_1 & I + \hat{b}_0 & \dots & \hat{b}_{-N+1} & \hat{b}_{-N} & \dots \\ \vdots & \vdots & & \vdots & \vdots & \\ \hat{b}_N & \hat{b}_{N-1} & \dots & I + \hat{b}_0 & \hat{b}_{-1} & \dots \end{pmatrix}$$

and thus $\|P_N T(a) - P_N\|_2^2$ equals

$$(N+1)(\|\hat{b}_0\|_2^2 + \|\hat{b}_{-1}\|_2^2 + \dots) + N\|\hat{b}_1\|_2^2 + \dots \|\hat{b}_N\|_2^2$$
$$\leq (N+1)\sum_{n \in \mathbb{Z}} \|\hat{b}_n\|_2^2 < \infty,$$

which shows that $P_N T(a) - P_N =: K$ and $P_N - P_N T^{-1}(a) = KT^{-1}(a)$ are in $\mathcal{C}_2(l^2(\mathcal{H}))$ and so gives the first inclusions in (6) and (7). The second inclusions of (6) and (7) follow from passage to adjoints.

(b) From (6) we infer that $P_N T(a)P_N - P_N$ and $P_N T(a)Q_N = (P_N T(a) - P_N)Q_N$ are in $\mathcal{C}_2(l^2(\mathcal{H}))$, and so (8) and (9) result immediately from the identity (4). Also by (4), we have

$$\begin{aligned} \Delta(a) &= P_0 T(a)P_0 - BQ_0(Q_0 T(a)Q_0)^{-1}Q_0 C \\ &= I + \hat{b}_0 - BQ_0(Q_0 T(a)Q_0)^{-1}Q_0 C \end{aligned}$$

with

$$\begin{aligned} B &= P_0 T(a)Q_0 = (P_0 T(a) - P_0)Q_0 \in \mathcal{C}_2(l^2(\mathcal{H})), \\ C &= Q_0 T(a)P_0 = Q_0(T(a)P_0 - P_0) \in \mathcal{C}_2(l^2(\mathcal{H})), \end{aligned}$$

and hence $BQ_0(Q_0 T(a)Q_0)^{-1}Q_0 C \in \mathcal{C}_1(l^2(\mathcal{H}))$. This proves (10).

(c) We have

$$P_N T(a) - P_N T(a_k) = P_N T(b) - P_N T(S_k b S_k) = P_N T(b) - \Lambda_k P_N T(b)\Lambda_k,$$

where $\Lambda_k : l^2(\mathcal{H}) \to l^2(\mathcal{H})$ is given by

$$\Lambda_k : \{f_0, f_1, \dots\} \mapsto \{S_k f_0, S_k f_1, \dots\}.$$

Because $\Lambda_k = \Lambda_k^* \to I$ strongly as $k \to \infty$ and $P_N T(b) = P_N T(a) - P_N$ is in $\mathcal{C}_2(l^2(\mathcal{H}))$ by (6), we may deduce the "+" version of (11) from Lemma 2.1. The "+" version of (12) then follows by taking adjoints. Further,

$$P_N T^{-1}(a_k) - P_N T^{-1}(a) = (P_N - P_N T(a_k))T^{-1}(a_k) - (P_N T^{-1}(a) - P_N).$$

By what has already been proved,

$$\|P_N - P_N T(a_k) - (P_N - P_N T(a))\|_2 \to 0 \quad \text{as} \quad k \to \infty,$$

and from Lemma 4.1 (with $p = \infty$) we know that

$$\|T^{-1}(a_k) - T^{-1}(a)\| \to 0 \quad \text{as} \quad k \to \infty.$$

Consequently, $P_N T^{-1}(a_k) - P_N T^{-1}(a)$ converges in the Hilbert-Schmidt norm to

$$(P_N - P_N T(a)) T^{-1}(a) - (P_N T^{-1}(a) - P_N) = 0,$$

which is equivalent to the "−" version of (11). The "−" version of (12) now results again from passage to adjoints.

Formula (4) shows that $\Delta_N(a_k)$ is equal to

$$P_N T(a_k) P_N - P_N T(a_k) Q_N (Q_N T(a_k) Q_N)^{-1} Q_N T(a_k) P_N,$$

and, by virtue of the "+" versions of (11) and (12) and Lemma 4.1 (with $p = \infty$) this goes in the Hilbert-Schmidt norm to

$$P_N T(a) P_N - P_N T(a) Q_N (Q_N T(a) Q_N)^{-1} Q_N T(a) P_N = \Delta_N(a). \qquad \blacksquare$$

We are now in a position to establish an operator-valued first Szegö-Widom limit theorem.

The determinant $\det (I + K)$ is well-defined if K is a trace class operator. If K is only a Hilbert-Schmidt operator, then $(I + K) e^{-K} - I$ is of trace class and one defines the so-called second regularized determinant $\det_2 (I + K)$ as

$$\det_2 (I + K) = \det ((I + K) e^{-K}).$$

Note that

$$\det_2 (I + K) = \det (I + K) e^{-\operatorname{tr} K}$$

whenever K is of trace class; here, of course, $\operatorname{tr} K$ stands for the trace of K. For several properties of the determinants $\det (I + K)$ and $\det_2 (I + K)$ we refer the reader to [3] and [7] and the literature quoted there.

THEOREM 4.3 *Let $a = I + b \in L^\infty(\mathcal{L}(\mathcal{H}))$, suppose $\mathcal{R}(b)$ is a compact subset of $\mathcal{C}_\infty(\mathcal{H})$, let $\hat{b}_n \in \mathcal{C}_2(\mathcal{H})$ for all $n \in \mathbb{Z}$ and $\sum_{n \in \mathbb{Z}} \|\hat{b}_n\|_2^2 < \infty$, and assume $T(a)$ is invertible and the finite section method is applicable to $T(a)$. Put $T_N(a) = P_N T(a) P_N \,|\, \operatorname{Im} P_N$. Then*

$$T_N(a) - I \in \mathcal{C}_2(\operatorname{Im} P_N) \quad \text{for all} \quad N \geq 0, \tag{14}$$

$$\Delta(a) - I \in \mathcal{C}_2(\mathcal{H}), \tag{15}$$

$$\Delta(a) - P_0 T(a) P_0 = -P_0 T(a) Q_0 (Q_0 T(a) Q_0)^{-1} Q_0 T(a) P_0 \in \mathcal{C}_1(\mathcal{H}) \tag{16}$$

and

$$\lim_{N \to \infty} \det_2 T_N(a) / \det_2 T_{N-1}(a) = G_2(a), \tag{17}$$

where

$$\begin{aligned}
G_2(a) &= (\det_2 \Delta(a)) \exp \operatorname{tr} (\Delta(a) - P_0 T(a) P_0) \tag{18} \\
&= (\det_2 \Delta(a)) \exp \operatorname{tr} (\Delta(a) - I - \hat{b}_0) \\
&= (\det_2 \Delta(a)) \exp \operatorname{tr} (- P_0 T(a) Q_0 (Q_0 T(a) Q_0)^{-1} Q_0 T(a) P_0).
\end{aligned}$$

Proof. The inclusions (14), (15), (16) follow from Lemma 4.2. Put $a_k = I + S_k b S_k$. Let $T_N(a)$ be invertible for all $N \geq N_0$ and define

$$R := \sup_{N \geq N_0} \| T_N^{-1}(a) P_N \|.$$

From Lemma 4.1 (with $p = \infty$) we deduce that there is a k_0 such that

$$\| T_N(a) P_N - T_N(a_k) P_N \| \leq \| T(a) - T(a_k) \| < 1/R$$

for all $k \geq k_0$. It follows that $T(a_k)$ and $T_N(a_k)$ are invertible for all $k \geq k_0$ and all $N \geq N_0$ and that

$$\sup_{k \geq k_0, N \geq N_0} \| T_N^{-1}(a_k) P_N \| \leq \infty.$$

Let now $k \geq k_0$ and $N \geq N_0 + 1$. We may interprete $T_N(a_k)$ and $T_{N-1}(a_k)$ as block Toeplitz matrices (with $(k+1) \times (k+1)$ blocks) and hence apply the Cramer-Jacobi rule to obtain that

$$\det T_{N-1}(a_k)/\det T_N(a_k) = \det P_0 T_N^{-1}(a_k) P_0.$$

Formula (5) shows that the right-hand side equals

$$\det \left(P_0 T^{-1}(a_k) P_0 - P_0 T^{-1}(a_k) Q_N (Q_N T^{-1}(a_k) Q_N)^{-1} Q_N T^{-1}(a_k) P_0 \right)$$
$$= \det A_k \det \left(I - \Delta(a_k) B_k D_{N,k} C_k \right),$$

where

$$A_k = P_0 T^{-1}(a_k) P_0 \,|\, \mathrm{Im}\, P_0, \qquad B_k = P_0 T^{-1}(a_k),$$
$$D_{N,k} = Q_N (Q_N T^{-1}(a_k) Q_N)^{-1} Q_N, \quad C_k = T^{-1}(a_k) P_0,$$

and we made use of the equality $A_k^{-1} = \Delta(a_k)$. We have

$$\det T_N(a_k) = \det{}_2 T_N(a_k) \exp \operatorname{tr} T_N(S_k b S_k) = \det{}_2 T_N(a_k) \exp \left((N+1) \operatorname{tr} S_k \hat{b}_0 S_k \right)$$

and

$$(\det A_k)^{-1} = \det \left(P_0 T^{-1}(a_k) P_0 \right)^{-1} P_0 = \det \Delta(a_k) = \det{}_2 \Delta(a_k) \exp \operatorname{tr} \left(\Delta(a_k) - P_0 \right),$$

with

$$\operatorname{tr} \left(\Delta(a_k) - P_0 \right) = \operatorname{tr} \left((P_0 T^{-1}(a_k) P_0)^{-1} P_0 - P_0 \right)$$
$$= \operatorname{tr} \left(P_0 T(a_k) P_0 - P_0 T(a_k) Q_0 (Q_0 T(a_k) Q_0)^{-1} Q_0 T(a_k) P_0 - P_0 \right)$$
$$= \operatorname{tr} \left(S_k \hat{b}_0 S_k - P_0 T(a_k) Q_0 (Q_0 T(a_k) Q_0)^{-1} Q_0 T(a_k) P_0 \right).$$

Hence

$$\det{}_2 T_{N-1}(a_k) \exp \left(N \operatorname{tr} S_k \hat{b}_0 S_k \right) / \left(\det{}_2 T_N(a_k) \exp \left((N+1) \operatorname{tr} S_k \hat{b}_0 S_k \right) \right)$$
$$= (\det{}_2 \Delta(a_k))^{-1} \exp \operatorname{tr} \left(- S_k \hat{b}_0 S_k + P_0 T(a_k) Q_0 (Q_0 T(a_k) Q_0)^{-1} Q_0 T(a_k) P_0 \right)$$
$$\times \det \left(I - \Delta(a_k) B_k D_{N,k} C_k \right)$$

and thus,

$$
\det{}_2 T_N(a_k)/\det{}_2 T_{N-1}(a_k)
$$
$$
= \det{}_2 \Delta(a_k) \exp \operatorname{tr} \left(- P_0 T(a_k) Q_0 (Q_0 T(a_k) Q_0)^{-1} Q_0 T(a_k) P_0 \right)
$$
$$
\times \det \left(I - \Delta(a_k) B_k D_{N,k} C_k \right)^{-1}.
$$

The limit passage $k \to \infty$ in the latter equality along with Lemmas 4.1 (for $p = \infty$) and 4.2 gives

$$
\det{}_2 T_N(a)/\det{}_2 T_{N-1}(a) \tag{19}
$$
$$
= \det{}_2 \Delta(a) \exp \operatorname{tr} \left(- P_0 T(a) Q_0 (Q_0 T(a) Q_0)^{-1} Q_0 T(a) P_0 \right)
$$
$$
\times \det \left(I - \Delta(a) P_0 T^{-1}(a) Q_N (Q_N T^{-1}(a) Q_N)^{-1} Q_N T^{-1}(a) P_0 \right)^{-1}
$$

for every $N \geq N_0 + 1$. We have

$$
D_N := (Q_N T^{-1}(a) Q_N)^{-1} Q_N
$$
$$
= Q_N T(a) Q_N - Q_N T(a) P_N (P_N T(a) P_N)^{-1} P_N T(a) Q_N,
$$

and since $Q_N \to 0$ and $(P_N T(a) P_N)^{-1} \to T^{-1}(a)$ strongly, it follows that $D_N \to 0$ strongly as $N \to \infty$. Lemma 4.2 tells us that $P_0 T^{-1}(a) = P_0 + K$ and $T^{-1}(a) P_0 = P_0 + L$ with $K, L \in \mathcal{C}_2(l^2(\mathcal{H}))$. Consequently,

$$
\| P_0 T^{-1}(a) Q_N (Q_N T^{-1}(a) Q_N)^{-1} Q_N T^{-1}(a) P_0 \|_1
$$
$$
= \| (P_0 + K) Q_N D_N Q_N (P_0 + L) \|_1 = \| K Q_N D_N Q_N L \|_1
$$
$$
= \| K D_N L \|_1 \leq \| K \|_2 \| D_N L \|_2
$$

and $\| D_N L \|_2 \to 0$ as $N \to \infty$ due to Lemma 2.1. Thus, the third factor on the right of (19) goes to 1 as $N \to \infty$, which implies (17) and (18). ∎

COROLLARY 4.4 *Let* $a = I + b \in L^\infty(\mathcal{L}(\mathcal{H}))$, *suppose* $\hat{b}_n \in \mathcal{C}_2(\mathcal{H})$ *for all* $n \in \mathbb{Z}$ *and*

$$
\sum_{n \in \mathbb{Z}} \| \hat{b}_n \| < \infty, \quad \sum_{n \in \mathbb{Z}} \| \hat{b}_n \|_2^2 < \infty. \tag{20}
$$

If both $T(a)$ *and* $T(\tilde{a})$ *are invertible, then (15) to (18) hold.*

Proof. We show that in the case at hand all hypotheses of Theorem 4.3 are satisfied.

Because $t^n \hat{b}_n$ is compact (even Hilbert-Schmidt) for all $t \in \mathbb{T}$ and all $n \in \mathbb{Z}$, and since $\sum_{n \in \mathbb{Z}} \| t^n \hat{b}_n \| < \infty$ for all $t \in \mathbb{T}$, it follows that b is a continuous function of $\mathbb{T}$ into $\mathcal{C}_\infty(\mathcal{H})$. The compactness of $\mathbb{T}$ so yields that $\mathcal{R}(b) = b(\mathbb{T})$ is a compact subset of $\mathcal{C}_\infty(\mathcal{H})$. Finally, we have $b \in C(\mathcal{C}_\infty(\mathcal{H}))$, and therefore Theorem 3.4 shows that the finite section method is applicable to $T(a)$. ∎

Here now is the Gohberg-Kaashoek version of the operator-valued first Szegö-Widom limit theorem.

COROLLARY 4.5 [6]. *Let* $a = I + b \in L^\infty(\mathcal{L}(\mathcal{H}))$, *suppose* $\hat{b}_n \in \mathcal{C}_2(\mathcal{H})$ *for all* $n \in \mathbb{Z}$ *and (20) is in force. If* $a(t)$ *is positive definite for each* $t \in \mathbb{T}$, *then (14) to (18) are valid.*

Proof. As the positive definiteness of a implies that both $T(a)$ and $T(\tilde{a})$ are invertible, the assertion is immediate from Corollary 4.4.   ∎

In conclusion we remark that Corollary 4.4 is no longer true if one drops the requirement that $T(\tilde{a})$ be invertible. Indeed, if we take $\mathcal{H} = \mathbb{C}^2$ and

$$a(t) = \begin{pmatrix} t & 1 \\ 0 & t^{-1} \end{pmatrix} \quad (t \in \mathbb{T}),$$

then $T(a)$ is invertible, $T(\tilde{a})$ is not invertible, and we have $\det_2 T_N(a) = 0$ for all $N \geq 0$ (see $[3, 7.19$ and $10.6]$).

5. The strong Szegö-Widom limit theorem.
Again we must start with a technical result.

LEMMA 5.1 *Let $a = I + b \in L^\infty(\mathcal{L}(\mathcal{H}))$, suppose $\mathcal{R}(b)$ is a compact subset of $C_\infty(\mathcal{H})$, and assume $T(a)$ and $T(\tilde{a})$ are invertible. Put $a_k = I + S_k b S_k$. Then a_k and $T(a_k^{-1})$ are invertible for all sufficiently large k and we have*

$$T_N(a_k) - I \in \mathcal{C}_1(\operatorname{Im} P_N), \tag{21}$$
$$P_N T^{-1}(a_k^{-1}) P_N - I \in \mathcal{C}_1(\operatorname{Im} P_N), \tag{22}$$
$$\det P_N T^{-1}(a_k^{-1}) P_N = (\det P_0 T^{-1}(a_k^{-1}) P_0)^{N+1} \tag{23}$$

for all $N \geq 0$.

Proof. Proposition 3.2 implies that a and $T(a^{-1})$ are invertible, and so Lemma 4.1 (with $p = \infty$) gives the invertibility of a_k and $T(a_k^{-1})$ for all k large enough. We clearly have

$$T_N(a_k) - P_N = T_N(S_k b S_k) \in \mathcal{C}_1(\operatorname{Im} P_N)$$

and thus (21) is true. Using (1) we get

$$T^{-1}(a_k^{-1}) = T(a_k) + T^{-1}(a_k^{-1}) H(a_k^{-1}) H(\tilde{a}_k), \tag{24}$$

and since obviously $H(\tilde{a}_k) P_N \in \mathcal{C}_1(\operatorname{Im} P_N)$, it follows that (22) is also valid.

To prove (23) we first identify $T^{-1}(a_k^{-1})$ and $T^{-1}(\tilde{a}_k)$ with infinite matrices $(A_{ij})_{i,j=0}^\infty$ and $(B_{ij})_{i,j=0}^\infty$, where A_{ij} and B_{ij} are $(k+1) \times (k+1)$ matrices, and then we put

$$X_N = (A_{ij})_{i,j=0}^N, \quad Y_N = (B_{ij})_{i,j=0}^N.$$

What we must show is that $\det X_N = (\det X_0)^{N+1}$.

Let $E_N = (0 \ \dots \ 0 \ I)^\mathsf{T}$ be the column constituted by N zero matrices of the size $(k+1) \times (k+1)$ and by the $(k+1) \times (k+1)$ identity matrix I. Then determine the column $F = (F_0 \dots F_N)^\mathsf{T}$ consisting of $N+1$ square matrices $F_0, \dots, F_N$ of the order $k+1$ as the solution of the equation $X_N F = E_N$. The Cramer-Jacobi rule gives that $\det F_N$ is equal to $\det X_{N-1}/\det X_N$. On the other hand, we have the identity

$$(P_N T^{-1}(a_k^{-1}) P_N)^{-1} P_N = W_N T^{-1}(\tilde{a}_k) W_N \tag{25}$$

(see [2, Lemma 6.12]), which in the notation employed here says that $X_N^{-1} = W_N Y_N W_N$, and hence

$$F = X_N^{-1} E_N = W_N Y_N W_N E_N = (B_{N0} \dots B_{00})^\mathsf{T}.$$

It follows that $\det X_{N-1}/\det X_N = \det B_{00}$ for all $N \geq 0$ and thus,

$$\det X_0/\det X_N = (\det B_{00})^N.$$

We finally have, again by (25),

$$\det B_{00} = \det W_0 T^{-1}(\tilde{a}_k) W_0 = \det (P_0 T^{-1}(a_k^{-1}) P_0)^{-1} P_0 = 1/\det X_0,$$

which gives that $\det X_N = (\det X_0)^{N+1}$. ∎

We define the $\mathcal{L}(\mathcal{H})$-valued Krein algebra $K(\mathcal{L}(\mathcal{H}))$ as the set of all $a = I+b \in L^\infty(\mathcal{L}(\mathcal{H}))$ such that

$$\hat{b}_n \in C_2(\mathcal{H}) \ \forall n \in \mathbb{Z} \quad \text{and} \quad \sum_{n \in \mathbb{Z}} |n| \, \| \hat{b}_n \|_2^2 < \infty. \tag{26}$$

It is easy to see that (26) is equivalent to the requirement that both $H(a)$ and $H(\tilde{a})$ be Hilbert-Schmidt on $l^2(\mathcal{H})$.

THEOREM 5.2 *The Krein algebra $K(\mathcal{L}(\mathcal{H}))$ is an algebra contained in the intersection of the two algebras (3). If $a \in K(\mathcal{L}(\mathcal{H}))$ is invertible in $L^\infty(\mathcal{L}(\mathcal{H}))$, then the inverse of a belongs to $K(\mathcal{L}(\mathcal{H}))$.*

Proof. See [2, Section 4.10]. ∎

The following theorem is one version of the operator-valued strong Szegő-Widom limit theorem.

THEOREM 5.3 *Let $a = I + b \in K(\mathcal{L}(\mathcal{H}))$, suppose $\mathcal{R}(b)$ is a compact subset of $C_\infty(\mathcal{H})$ and $T(a)$ and $T(\tilde{a})$ (and thus also a and $T(a^{-1})$) are invertible. Then*

$$T_N(a) - I \in C_2(\operatorname{Im} P_N), \tag{27}$$
$$\Delta^{-1}(a^{-1}) - I := P_0 T^{-1}(a^{-1}) P_0 - I \in C_2(\mathcal{H}), \tag{28}$$
$$P_0(T^{-1}(a^{-1}) - T(a)) P_0 \in C_1(\mathcal{H}), \tag{29}$$
$$T(a^{-1}) T(a) - I \in C_1(l^2(\mathcal{H})) \tag{30}$$

and we have

$$\lim_{N \to \infty} \det_2 T_N(a)/\Gamma_2(a)^{N+1} = E(a), \tag{31}$$

where

$$\Gamma_2(a) = (\det_2 \Delta^{-1}(a^{-1})) \exp \operatorname{tr} P_0(T^{-1}(a^{-1}) - T(a)) P_0, \tag{32}$$
$$E(a) = \det T(a^{-1}) T(a). \tag{33}$$

Proof. Inclusion (27) is immediate from (6). We have

$$T(a^{-1})T(a) = I - H(a^{-1})H(\tilde{a}), \tag{34}$$

and since $H(a^{-1})$ and $H(\tilde{a})$ are in $C_2(l^2(\mathcal{H}))$ by the definition of $K(\mathcal{L}(\mathcal{H}))$ and Theorem 5.2, we arrive at (30). Further, from (34) we obtain that

$$T^{-1}(a^{-1}) - T(a) = T^{-1}(a^{-1})H(a^{-1})H(\tilde{a}) \in C_2(l^2(\mathcal{H})),$$

which shows that (29) also holds. Because $P_0 T(a) P_0 = P_0 + \hat{b}_0$ and $\hat{b}_0 \in C_2(\mathcal{H})$, (28) is a consequence of (29).

From (11) we deduce that $\det_2 T_N(a) = \lim_{k \to \infty} \det_2 T_N(a_k)$, and in view of (21) we have

$$\det_2 T_N(a_k) = \det T_N(a_k) \exp \operatorname{tr} (P_N - T_N(a_k)).$$

Recalling that $P_N T^{-1}(a_k^{-1}) P_N \,|\, \operatorname{Im} P_N = \Delta_N^{-1}(a_k^{-1})$ and taking into account (24) we get

$$T_N(a_k) = \Delta_N^{-1}(a_k^{-1})(P_N - \Delta_N(a_k^{-1})P_N T^{-1}(a_k^{-1})H(a_k^{-1})H(\tilde{a}_k)P_N).$$

Since $a \in K(\mathcal{L}(\mathcal{H}))$, the operator $H(\tilde{a}_k)$ is Hilbert-Schmidt and

$$\|H(\tilde{a}) - H(\tilde{a}_k)\|_2 = \|H(\tilde{a}) - \Lambda_k H(\tilde{a})\Lambda_k\|_2 \to 0 \quad \text{as} \quad k \to \infty \tag{35}$$

by Lemma 2.1. From (2) we obtain

$$H(a_k^{-1}) = -T^{-1}(a_k)H(a_k)T(\tilde{a}_k^{-1}),$$

hence $H(a_k^{-1}) \in C_2(l^2(\mathcal{H}))$ and thus, by (35),

$$\|H(a^{-1}) - H(a_k^{-1})\|_2 \to 0 \quad \text{as} \quad k \to \infty. \tag{36}$$

Consequently, as $k \to \infty$,

$$\det (P_N - \Delta_N(a_k^{-1})P_N T^{-1}(a_k^{-1})H(a_k^{-1})H(\tilde{a}_k)P_N)$$
$$\to \det (P_N - \Delta_N(a^{-1})P_N T^{-1}(a^{-1})H(a^{-1})H(\tilde{a})P_N) =: E_N(a).$$

Next, by Lemma 5.1,

$$\det \Delta_N^{-1}(a_k^{-1}) \exp \operatorname{tr} (P_N - T_N(a_k))$$
$$= (\det \Delta_0^{-1}(a_k^{-1}))^{N+1} \exp \operatorname{tr} (P_N - T_N(a_k))$$
$$= (\det \Delta_0^{-1}(a_k^{-1}) \exp \operatorname{tr} (P_0 - T_0(a_k))^{N+1}$$
$$= (\det_2 \Delta_0^{-1}(a_k^{-1}) \exp \operatorname{tr} (P_0 T^{-1}(a_k^{-1})P_0 - P_0 + P_0 - T_0(a_k))^{N+1}$$
$$= (\det_2 \Delta_0^{-1}(a_k^{-1}) \exp \operatorname{tr} P_0(T^{-1}(a_k^{-1}) - T(a_k))P_0)^{N+1}$$

(recall that $\Delta_0 =: \Delta$). Combining (6), (24), (35), (36) we see that

$$\det_2 \Delta^{-1}(a_k^{-1}) = \det_2 P_0 T^{-1}(a_k^{-1})P_0 \to \det_2 P_0 T^{-1}(a^{-1})P_0 = \det_2 \Delta^{-1}(a^{-1}), \tag{37}$$
$$\operatorname{tr} P_0(T^{-1}(a_k^{-1}) - T(a_k))P_0 \to \operatorname{tr} P_0 (T^{-1}(a^{-1}) - T(a))P_0 \tag{38}$$

as $k \to \infty$. Thus, what we have proved is that

$$\det_2 T_N(a) = \Gamma_2(a)^{N+1} E_N(a),$$

where $\Gamma_2(a)$ is given by (32). Since

$$\Delta_N(a^{-1}) = (P_N T^{-1}(a^{-1}) P_N)^{-1} \to (T^{-1}(a^{-1}))^{-1} = T(a^{-1}) \quad \text{strongly}$$

by Theorem 3.3, we finally arrive at the equality

$$\lim_{N \to \infty} E_N(a) = \det (I - T(a^{-1}) T^{-1}(a^{-1}) H(a^{-1}) H(\tilde{a}))$$
$$= \det (I - H(a^{-1}) H(\tilde{a})) = \det T(a^{-1}) T(a) = E(a),$$

which gives (31) and (32). ∎

The following proposition reveals the connection between Theorems 5.3 and 4.3.

PROPOSITION 5.4 *If $a \in L^\infty(\mathcal{L}(\mathcal{H}))$ is subject to the hypotheses of Theorem 5.3, then a also satisfies the hypotheses of Theorem 4.3 and we have $\Gamma_2(a) = G_2(a)$.*

Proof. The assertion that the hypotheses of Theorem 5.3 are stronger than those of Theorem 4.3 follows from Theorem 3.4 and the fact that $K(\mathcal{L}(\mathcal{H}))$ is a subset of $C(\mathcal{C}_\infty(\mathcal{H})) + H^\infty(\mathcal{L}(\mathcal{H}))$.

Let us show the equality $\Gamma_2(a) = G_2(a)$. We know from (13) that

$$\|\Delta(a_k) - \Delta(a)\|_2 \to 0 \quad \text{as} \quad k \to \infty, \tag{39}$$

and since, by (4),

$$(P_0 T^{-1}(a_k) P_0)^{-1} P_0 - P_0 T(a_k) P_0 = -P_0 T(a_k) Q_0 (Q_0 T(a_k) Q_0)^{-1} Q_0 T(a_k) P_0$$

and, by Lemmas 4.1 (with $p = \infty$) and 4.2,

$$\|P_0 T(a_k) - P_0 T(a)\|_2 \to 0, \quad \|T(a_k) P_0 - T(a) P_0\|_2 \to 0,$$

$$\|(Q_0 T(a_k) Q_0)^{-1} Q_0 - (Q_0 T(a) Q_0)^{-1} Q_0\| \to 0$$

as $k \to \infty$, it follows that

$$\text{tr} (\Delta(a_k) - P_0 T(a_k) P_0) \to \text{tr} (\Delta(a) - P_0 T(a) P_0) \quad \text{as} \quad k \to \infty. \tag{40}$$

The relations (37) to (40) show that it is enough to prove that $\Gamma_2(a_k) = G_2(a_k)$. We have

$$G_2(a_k) = \det{}_2 \Delta(a_k) \exp \text{tr} (\Delta(a_k) - P_0 T(a_k) P_0)$$
$$= \det \Delta(a_k) \exp \text{tr} (P_0 - P_0 T(a_k) P_0),$$
$$\Gamma_2(a_k) = \det{}_2 \Delta^{-1}(a_k^{-1}) \exp \text{tr} (\Delta^{-1}(a_k^{-1}) - P_0 T(a_k) P_0)$$
$$= \det \Delta^{-1}(a_k^{-1}) \exp \text{tr} (P_0 - P_0 T(a_k) P_0),$$

and hence we are left with showing that

$$\det P_0 T^{-1}(a_k^{-1}) P_0 = \det (P_0 T^{-1}(a_k) P_0)^{-1} P_0,$$

or, owing to the identity (25), that

$$\det P_0 T^{-1}(a_k^{-1}) P_0 = \det P_0 T^{-1}(\tilde{a}_k) P_0.$$

But

$$\det P_0 T^{-1}(\tilde{a}_k) P_0$$
$$= \lim_{N\to\infty} P_0 T_N^{-1}(\tilde{a}_k) P_0 \qquad\qquad (Theorems\ 5.2\ and\ 3.4)$$
$$= \lim_{N\to\infty} \det T_{N-1}(\tilde{a}_k)/\det T_N(\tilde{a}_k) \qquad (Cramer - Jacobi)$$
$$= \lim_{N\to\infty} \det W_{N-1} T_{N-1}(a_k) W_{N-1}/\det W_N T_N(a_k) W_N$$
$$= \lim_{N\to\infty} \det T_{N-1}(a_k)/\det T_N(a_k)$$
$$= \lim_{N\to\infty} \det P_0 T_N^{-1}(a_k) P_0 \qquad\qquad (Cramer - Jacobi)$$
$$= \det P_0 T^{-1}(a_k) P_0 \qquad\qquad (Theorems\ 5.2\ and\ 3.4). \qquad \blacksquare$$

We remark that if $a \in L^\infty(\mathcal{L}(\mathcal{H}))$ satisfies the hypotheses of Corollary 4.5 or Theorem 5.3 and, in addition, $\hat{b}_0 \in \mathcal{C}_1(\mathcal{H})$, then $\Delta(a) - I \in \mathcal{C}_1(\mathcal{H})$ in view of (10), and so $G_2(a)$ or $\Gamma_2(a)$ may be replaced by

$$\det_2 \Delta(a) \exp \operatorname{tr}(\Delta(a) - I - \hat{b}_0) = \det \Delta(a) \exp \operatorname{tr}(-\hat{b}_0).$$

In case $\hat{b}_n \in \mathcal{C}_1(\mathcal{H})$ for all $n \in \mathbb{Z}$ we have the following.

COROLLARY 5.5 *In addition to the hypotheses of Theorem 5.3 suppose that* $\hat{b}_n \in \mathcal{C}_1(\mathcal{H})$ *for all* $n \in \mathbb{Z}$. *Then*

$$T_N(a) - I \in \mathcal{C}_1(\operatorname{Im} P_N), \Delta(a) - I \in \mathcal{C}_1(\mathcal{H})$$

and

$$\lim_{N\to\infty} \det T_N(a)/G_1(a)^{N+1} = E(a),$$

where $G_1(a) = \det \Delta(a)$ *and* $E(a) = \det T(a^{-1})T(a)$.

Proof. Clearly $T_N(a) - I \in \mathcal{C}_1(\operatorname{Im} P_N)$, and we have already observed that $\Delta(a) - I \in \mathcal{C}_1(\mathcal{H})$ whenever $\hat{b}_0 \in \mathcal{C}_1(\mathcal{H})$. Since

$$\det_2 T_N(a) = \det T_N(a) \exp \operatorname{tr}(T_N(a) - P_N)$$
$$= \det T_N(a) \exp \operatorname{tr}(-(N+1)\hat{b}_0)$$

and $\Gamma_2(a) = \det \Delta(a) \exp \operatorname{tr}(-\hat{b}_0)$, the assertion is so immediate from Theorem 5.3. $\qquad \blacksquare$

Let us finally state a result for positive definite symbols.

COROLLARY 5.6 *Let* $p \in \{1, 2\}$. *Suppose* $a = I + b \in K(\mathcal{L}(\mathcal{H}))$ *and* $\mathcal{R}(b)$ *is a compact subset of* $\mathcal{C}_\infty(\mathcal{H})$. *In case* $p = 1$, *assume also that* $\hat{b}_n \in \mathcal{C}_1(\mathcal{H})$ *for all* $n \in \mathbb{Z}$. *Then if all operators in* $\mathcal{R}(a)$ *are positive definite, we have*

$$\lim_{N\to\infty} \det_p T_N(a)/G_p(a)^{N+1} = E(a),$$

where $G_2(a)$ *is given by (18), while* $G_1(a)$ *and* $E(a)$ *are as in Corollary 5.5.*

Proof. It suffices to note that the positive definiteness of a implies the invertibility of both $T(a)$ and $T(\tilde{a})$. ∎

6. Continuous symbols. In what follows we let $C(\mathcal{L}(\mathcal{H}))$ and $C(\mathcal{C}_p(\mathcal{H}))$ ($p = 1, 2, \infty$) stand for the continuous functions of $\mathbb{T}$ into $\mathcal{L}(\mathcal{H})$ and $\mathcal{C}_p(\mathcal{H})$,respectively. For $b \in C(\mathcal{C}_p(\mathcal{H}))$, we define

$$\|b\|_p = \max_{t \in \mathbb{T}} \|b(t)\|_p.$$

Note that if $b \in C(\mathcal{C}_p(\mathcal{H}))$, then $\mathcal{R}(b) = b(\mathbb{T})$ is a compact subset of $C(\mathcal{C}_p(\mathcal{H}))$.

Our first concern is index formulas for Fredholm Toeplitz operators with symbols in $I + C(\mathcal{C}_p(\mathcal{H}))$. If $c : \mathbb{T} \to \mathbb{C}$ is a continuous function without zeros, we denote by wind c the winding number with respect to the origin of the naturally oriented curve $c(\mathbb{T})$.

PROPOSITION 6.1 *Let $a = I + b$ with $b \in C(\mathcal{C}_p(\mathcal{H}))$, where $p \in \{1, 2, \infty\}$. Then $T(a)$ is Fredholm on $l^2(\mathcal{H})$ if and only if a is invertible in $L^\infty(\mathcal{L}(\mathcal{H}))$ and $a^{-1} = I + c$ with $c \in C(\mathcal{C}_p(\mathcal{H}))$. In that case $\det(S_k + S_k b S_k)$ does not vanish on $\mathbb{T}$ for all sufficiently large k and the index of $T(a)$ is given by*

$$\operatorname{Ind} T(a) = -\lim_{k \to \infty} \operatorname{wind} \det(S_k + S_k b S_k).$$

Proof. The "if" portion of the Fredholm criterion follows from Theorem 2.2. The Hartman-Wintner theorem shows that a is invertible in $L^\infty(\mathcal{L}(\mathcal{H}))$ whenever $T(a)$ is Fredholm. Let $a^{-1} = I + c$. Then $(I + b)(I + c) = I$, whence $c = -b - bc$, which yields that c is in $C(\mathcal{C}_p(\mathcal{H}))$.

From Lemma 4.1 (with $p = \infty$) we infer that

$$\operatorname{Ind} T(a) = \lim_{k \to \infty} \operatorname{Ind} T(I + S_k b S_k) = \lim_{k \to \infty} \operatorname{Ind} T(S_k + S_k b S_k),$$

where $T(S_k + S_k b S_k)$ is thought of as a block Toeplitz operator acting on $l^2(S_k \mathcal{H})$. We so have

$$\operatorname{Ind} T(S_k + S_k b S_k) = -\operatorname{wind} \det(S_k + S_k b S_k). \quad ∎$$

If $a = I + b$ with $b \in C(\mathcal{C}_p(\mathcal{H}))$ and $p \in \{1, 2\}$, then the function

$$\det_p a : \mathbb{T} \to \mathbb{C}, \quad t \mapsto \det_p(I + b(t))$$

is continuous. Note that $\det_p a(t) \neq 0$ for all $t \in \mathbb{T}$ if and only if a is invertible in $L^\infty(\mathcal{L}(\mathcal{H}))$ (and thus in $I + C(\mathcal{C}_p(\mathcal{H}))$).

COROLLARY 6.2 *Let $a = I + b$ with $b \in C(\mathcal{C}_p(\mathcal{H}))$ and $p \in \{1, 2\}$. If $T(a)$ is Fredholm, then*

$$\operatorname{Ind} T(a) = -\operatorname{wind} \det_p a.$$

Proof. If $K, L \in C_p$, then

$$|\det_p (I + K) - \det_p (I + L)| \leq \|K - L\|_p \exp (\Gamma_p(\|K\|_p + \|L\|_p + 1)^p),$$

where $\Gamma_1 = 1$ and $\Gamma_2 = 1/2$ (see e.g. [3, 1.8(c)]). We so infer from Lemma 4.1 that

$$\max_{t \in \mathbb{T}} |\det_p (S_k + S_k b(t) S_k) - \det_p (I + b(t))| \tag{41}$$
$$= \max_{t \in \mathbb{T}} |\det_p (I + S_k b(t) S_k) - \det_p (I + b(t))|$$
$$\leq \|b - S_k b S_k\|_p \exp (2 \|b\|_p + 1)^2 = o(1) \quad (k \to \infty).$$

For $p = 1$ the assertion of the corollary is thus immediate from Proposition 6.1. Since

$$\text{wind} \det_2 (I + S_k b S_k) = \text{wind} (\det (I + S_k b S_k) \exp \text{tr} (-S_k b S_k))$$
$$= \text{wind} \det (I + S_k b S_k) + \text{wind} \exp \text{tr} (-S_k b S_k)$$
$$= \text{wind} \det (I + S_k b S_k) = \text{wind} \det (S_k + S_k b S_k),$$

we obtain the corollary also for $p = 2$. $\blacksquare$

We remark that the preceding corollary can be extended to all p between 1 and ∞.

We are now in a position to give alternative expressions for the terms $G_2(a)$ and $G_1(a)$ appearing in Theorem 4.3 and Corollary 5.5.

PROPOSITION 6.3 *Let $a = I + b$ be in $L^\infty(\mathcal{L}(\mathcal{H}))$, suppose $T(a)$ is invertible, and assume that $\hat{b}_n \in C_2(\mathcal{H})$ for all $n \in \mathbb{Z}$. If $b \in C(C_1(\mathcal{H}))$ then $\Delta(a) - I \in C_1(\mathcal{H})$ and*

$$G_1(a) := \det \Delta(a) = \exp \frac{1}{2\pi} \int_0^{2\pi} \log \det a(e^{i\theta}) \, d\theta, \tag{42}$$

and if $b \in C(C_2(\mathcal{H}))$ then

$$\Delta(a) - I \in C_2(\mathcal{H}), \quad \Delta(a) - I - \hat{b}_0 \in C_1(\mathcal{H}) \tag{43}$$

and

$$G_2(a) := \det_2 \Delta(a) \exp \text{tr} (\Delta(a) - I - \hat{b}_0)$$
$$= \exp \frac{1}{2\pi} \int_0^{2\pi} \log \det_2 a(e^{i\theta}) \, d\theta. \tag{44}$$

Proof. We first remark that Proposition 6.1 and Corollary 6.2 imply that the logarithms in (42) and (44) are well-defined.

Let first $b \in C(C_1(\mathcal{H}))$. Then

$$\hat{b}_0 = \frac{1}{2\pi} \int_0^{2\pi} b(e^{i\theta}) \, d\theta \in C_1(\mathcal{H})$$

and hence $\Delta(a) - I \in C_1(\mathcal{H})$ due to (10). Put $a_k = I + S_k b S_k$. As in the proof of Lemma 4.2, we have

$$\Delta(a_k) - I = S_k \hat{b}_0 S_k - P_0 T(a_k) Q_0 (Q_0 T(a_k) Q_0)^{-1} Q_0 T(a_k) P_0,$$

and since $a_k \to a$ in $L^\infty(\mathcal{L}(\mathcal{H}))$ (Lemma 4.1 with $p = \infty$), $S_k \hat{b}_0 S_k \to \hat{b}_0$ in $\mathcal{C}_1(\mathcal{H})$ (Lemma 2.1 with $p = 1$), we deduce from the "+" versions of (11) and (12) (with $N = 0$) that $\Delta(a_k) \to \Delta(a)$ in $\mathcal{C}_1(\mathcal{H})$, which implies that

$$G_1(a) = \lim_{k \to \infty} \det \Delta(a_k).$$

From (41) we know that $\det a_k \to \det a$ uniformly on $\mathbb{T}$, and hence it suffices to show (42) with a replaced by a_k. In this case, however, (42) is well known (see [11] or [2, 6.6]).

Now let $b \in C(\mathcal{C}_2(\mathcal{H}))$. Then (43) is (9) and (10), and (13) gives us that $\Delta(a_k) \to \Delta(a)$ in $\mathcal{C}_2(\mathcal{H})$. Moreover, as above we may conclude that

$$\Delta(a_k) - I - S_k \hat{b}_0 S_k \to \Delta(a) - I - \hat{b}_0 \text{ in } \mathcal{C}_1(\mathcal{H}).$$

Hence $G_2(a_k) \to G_2(a)$. Since, by (41), $\det_2 a_k \to \det_2 a$ uniformly on $\mathbb{T}$, it is again enough to verify (44) with a_k in place of a. But in that case we have

$$\det_2 \Delta(a_k) \exp \operatorname{tr} \left(\Delta(a_k) - I - S_k \hat{b}_0 S_k \right)$$

$$= \det \Delta(a_k) \exp \operatorname{tr} \left(-S_k \hat{b}_0 S_k \right)$$

$$= \exp \frac{1}{2\pi} \int_0^{2\pi} \log \det a_k(e^{i\theta})\, d\theta \ \exp \operatorname{tr} \left(-S_k \hat{b}_0 S_k \right)$$

$$= \exp \frac{1}{2\pi} \int_0^{2\pi} \log \det_2 a_k(e^{i\theta})\, d\theta \ \exp \frac{1}{2\pi} \int_0^{2\pi} \log \exp \operatorname{tr} \left(a_k(e^{i\theta}) - I \right) d\theta$$

$$\times \exp \operatorname{tr} \left(-S_k \hat{b}_0 S_k \right)$$

$$= \exp \frac{1}{2\pi} \int_0^{2\pi} \log \det_2 a_k(e^{i\theta})\, d\theta \ \exp \frac{1}{2\pi} \operatorname{tr} S_k \int_0^{2\pi} b(e^{i\theta})\, d\theta \ S_k$$

$$\times \exp \operatorname{tr} \left(-S_k \hat{b}_0 S_k \right)$$

$$= \exp \frac{1}{2\pi} \int_0^{2\pi} \log \det_2 a_k(e^{i\theta})\, d\theta. \quad \blacksquare$$

The integral terms in (42) and (44) also make sense if the assumption that $T(a)$ be invertible (which enters essentially into the definition of $\Delta(a)$) is replaced by the requirement that $T(a)$ be merely Fredholm of index zero. Because checking the latter is much simpler than convincing oneself whether $T(a)$ and $T(\tilde{a})$ are invertible, the following theorem, which under slightly weaker smoothness hypotheses was established by Widom [11] in the case $\dim \mathcal{H} < \infty$, is of considerable interest.

THEOREM 6.4 *Let* $a = I + b \in K(\mathcal{L}(\mathcal{H}))$ *and* $p \in \{1, 2\}$. *Suppose* $b \in C(\mathcal{C}_p(\mathcal{H}))$, *and in case* $p = 1$, *assume also that* $\hat{b}_n \in \mathcal{C}_1(\mathcal{H})$ *for all* $n \in \mathbb{Z}$. *If* $T(a)$ *is Fredholm of index zero on* $l^2(\mathcal{H})$, *then*

$$T_N(a) - I \in \mathcal{C}_p(\operatorname{Im} P_N), \quad T(a^{-1})T(a) - I \in \mathcal{C}_1(l^2(\mathcal{H}))$$

and

$$\lim_{N \to \infty} \det_p T_N(a)/G_p(a)^{N+1} = E(a)$$

with

$$G_p(a) = \exp \frac{1}{2\pi} \int_0^{2\pi} \log \det_p a(e^{i\theta})\, d\theta, \quad E(a) = \det T(a^{-1})T(a).$$

Proof. This can be proved with the help of a trick by Widom [10], [11] (also see [2, 6.14]).

First one can show that if f and g are two nonzero elements of $l^2(\mathcal{H})$, then there is a trigonometric polynomial $c : \mathbb{T} \to \mathcal{L}(\mathcal{H})$ such that $(T(c)f, g) \neq 0$. To see this, choose any orthonormal basis $\{e_n\}_{n=0}^{\infty}$ in $\mathcal{H}$ and re-interpret $l^2(\mathcal{H})$ as the Hilbert space of all $\mathcal{H}$-valued functions

$$h : \mathbb{T} \to \mathcal{H}, \quad h(t) = \sum_{n=0}^{\infty} h_n(t) e_n \quad (t \in \mathbb{T})$$

such that h_n belongs to the Hardy space $H^2(\mathbb{T})$ for every n and

$$\|h\|^2 := \sum_{n=0}^{\infty} \|h_n\|_{H^2(\mathbb{T})}^2 < \infty.$$

If now f and g are nonzero elements of $l^2(\mathcal{H})$, then there are j_0 and i_0 such that $f_{j_0} \in H^2(\mathbb{T})$ and $g_{i_0} \in H^2(\mathbb{T})$ are nonzero. It follows that there is a trigonometric polynomial $c_{i_0 j_0} : \mathbb{T} \to \mathbb{C}$ such that $\int c_{i_0 j_0} f_{j_0} g_{i_0} \neq 0$, and if we define $c(t) \in \mathcal{L}(\mathcal{H})$ for $t \in \mathbb{T}$ as the operator whose i_0, j_0 entry in its matrix representation with respect to the basis $\{e_n\}$ is $c_{i_0 j_0}(t)$ and all other entries of which are zero, then $(T(c)f, g) \neq 0$ (see [10, p.30]). Notice that the c so constructed belongs to $C(\mathcal{C}_1(\mathcal{H}))$ and that the values of c as well as its Fourier coefficients are actually finite-rank operators.

The theorem of [10] in conjunction with what was said in the preceding paragraph implies that one can find $d = I + c \in K(\mathcal{L}(\mathcal{H}))$ such that $c \in C(\mathcal{C}_p(\mathcal{H}))$, $\hat{c}_n \in \mathcal{C}_1(\mathcal{H})$ for all $n \in \mathbb{Z}$ whenever $p = 1$, both $T(d)$ and $T(\tilde{d})$ are invertible, and $T(a_\zeta) := T((1 - \zeta)a + \zeta d)$ is Fredholm of index zero for all ζ in an open neighborhood F of the disk $|\zeta| \leq 1$ (see [11] or [2, 6.14]).

The proof may now be finished as in [11] or [2, 6.14]: the function $E(\zeta) = \det T(a_\zeta^{-1})T(a_\zeta)$ is analytic in F and $E(1) \neq 0$; so there is an $r \in (0, 1)$ such that $E(\zeta) \neq 0$ for $|\zeta| = r$ and consequently, $T(a_\zeta)$ and $T(\tilde{a}_\zeta)$ are invertible for $|\zeta| = r$; Theorem 5.3 resp. Corollary 5.5 along with Proposition 6.3 imply that

$$\lim_{N \to \infty} \det_p T_N(a_\zeta) \exp\left(-\frac{N+1}{2\pi} \int_0^{2\pi} \log \det_p a_\zeta(e^{i\theta}) \, d\theta\right) = E(a_\zeta)$$

for $|\zeta| = r$; this holds uniformly on the circle $|\zeta| = 1$, thus throughout the disk $|\zeta| \leq 1$ and in particular at $\zeta = 0$. ∎

References

[1] H. Bart, I. Gohberg, and M.A. Kaashoek: The coupling method for solving integral equations. *Operator Theory: Advances and Applications* **12** (1984), 39-73.

[2] A. Böttcher and B. Silbermann: *Invertibility and Asymptotics of Toeplitz Matrices.* Akademie-Verlag, Berlin (1983).

[3] A. Böttcher and B. Silbermann: *Analysis of Toeplitz Operators.* Akademie-Verlag, Berlin (1989) and Springer-Verlag, Berlin, Heidelberg, New York (1990).

[4] A. Devinatz and M. Shinbrot: General Wiener-Hopf operators. *Trans. Amer. Math. Soc.* **145** (1969), 467-494.

[5] I. Gohberg and I. Feldman: *Convolution Equations and Projection Methods for Their Solution.* Amer. Math. Soc. Transl. of Math. Monographs **41**, Providence, R. I, 1974.

[6] I. Gohberg and M.A. Kaashoek: Asymptotic formulas of Szegö-Kac-Achiezer type. *Asymptotic Analysis* **5** (1992), 187-220.

[7] I. Gohberg and M.G. Krein: *Introduction to the Theory of Linear Non-selfadjoint Operators in Hilbert Space.* Amer. Math. Soc. Transl. of Math. Monographs **18**, Providence, R. I., 1969.

[8] L.B. Page, Bounded and compact vectorial Hankel operators, *Trans. Amer. Math. Soc.* **150** (1970), 529–539.

[9] F.-O. Speck: *General Wiener-Hopf Factorization Methods.* Pitman Research Notes **119**, Pitman, Boston, London, Melbourne 1985.

[10] H. Widom, Perturbing Fredholm operators to obtain invertible operators, *J. Funct. Anal.* **20** (1975), 26–31.

[11] H. Widom, Asymptotic behavior of block Toeplitz matrices and determinants II, *Adv. Math.* **21** (1976), 1–29.

Technische Universität Chemnitz
Fachbereich Mathematik
PSF 964
D-09009 Chemnitz
Germany

AMS subject classification: 47B35

Operator Theory:
Advances and Applications, Vol. 71
© 1994 Birkhäuser Verlag Basel/Switzerland

THE ADAMJAN-AROV-KREIN THEOREM IN GENERAL AND REGULAR REPRESENTATIONS OF R² AND THE SYMPLECTIC PLANE

Mischa Cotlar and Cora Sadosky[1]

Dedicated to our friend Harold Widom, in his 60th birthday.

The classical Adamjan-Arov-Krein theorem relating the singular numbers of Hankel operators to best approximations of their symbols has an abstract version for unitary representations of the discrete group $\mathbb{Z}^n$, $n \geq 1$ [CS4]. Here we obtain similar results for the regular representations of $\mathbb{R}^n$, $n \geq 1$, and the symplectic spaces. This is based on lifting theorems and on the characterizations of the generators and cogenerators of semiunitary semigroups. In particular, it is shown that all compact Hankel operators in $\mathbb{R}^n$, $n \geq 1$, and in the symplectic spaces are zero.

INTRODUCTION. Analogues of the Adamjan-Arov-Krein (A-A-K) theorem [AAK] were given in [CS4] for Hankel forms in general and regular representations of the groups $\mathbb{Z}^n$, $n \geq 1$. Here we show that, using the

---

[1]Author partially supported by NSF(USA) grant DMS89-11717.

well-known characterizations of the **generators** and **cogenerators** of one-parameter semiunitary semigroups, results similar to those in [CS4] can be proved both for continuous unitary representations of $\mathbb{R}^n$, and of the symplectic spaces $(\mathbb{R}^{2n}, [,])$ $n \geq 1$. This presentation is self-contained, all definitions and results from [CS4] being summarized below. For notational simplicity, we detail only the bidimensional cases.

The main difference between the results in the discrete case treated in [CS4] and those in one of the continuous cases treated here is in the nature of the symbols of the Hankel forms.

In all bidimensional cases, two kind of such symbols can be given. In the discrete case of $\mathbb{Z}^2$, the first kind consists in pairs of bounded functions defined in $\mathbf{T}^2$, while the second kind is given by pairs of operator-valued functions defined in $\mathbf{T}$. As we show below, a similar situation occurs in the continuous case of $\mathbb{R}^2$. However, in the case of the symplectic plane, this is not so, since the symbols of the first kind are pairs of bounded operators acting in $L^2(\mathbb{R})$, instead of bounded functions. This gives the corresponding Hankel forms in terms of pairs of symbols that are either Hilbert-Schmidt operators or operator-valued functions, which, through a Weyl isomorphism, are shown to satisfying A-A-K conditions in terms of Blaschke-Potapov products.

The study of Hankel forms in $\mathbb{R}$ is reduced to that of those in $\mathbb{Z}$ through the consideration of the cogenerators of the corresponding semigroups. However such reduction is not sufficient to deal with the problem in $\mathbb{R}^2$, where the forms are required to be invariant under two shifts.

The organization of the paper is as follows. The reduction of the Hankel forms in $\mathbb{R}$ to those in $\mathbb{Z}$ is done in Section 1, leading to versions of the A-A-K theorem for both the general and regular representation of $\mathbb{R}$, given in Section 2. In Section 3, the abstract A-A-K theorem for general representations of $\mathbb{R}^2$ and $(\mathbb{R}^2, [,])$ is proved. In Section 4 a more precise form of the theorem is given for the case of the regular representation of $\mathbb{R}^2$, through the two kinds of symbols referred above. A similar, although more involved, study of the different pairs of symbols in the case of the regular representation of the symplectic plane is done in Section 5.

Only some of the continuous analogues of the results in [CS4] are developed here. Other results, as those on weighted Hardy spaces and on

Sarason commutants, can also be given continuous versions through similar procedure.

A more detailed study of the $(\mathbb{R}^{2n}, [\ , \ ])$ case in connection with Fock spaces, as well as of the forms invariant with respect to nonunitary evolution semigroups, will be the object of a forthcoming paper.

Both authors enjoyed the hospitality of the Institut Mittag-Leffler in 1990-91 while pursuing the work reported here and in [CS4]. It is our pleasure to acknowledge the Institut's support as well as to thank both its faculty and staff for making our visits so pleasant.

1. HANKEL FORMS IN DISCRETE AND CONTINUOUS EVOLUTION SYSTEMS.

Given a Hilbert space H and a discrete group of unitary operators in H, $\{\tau^n, \ n \ \epsilon \ \mathbb{Z}\}$ (i.e., an unitary representation of the group $\mathbb{Z}$), $[H; \ \tau^n, \ n \ \epsilon \ \mathbb{Z}]$ is called a _discrete evolution system_. Similarly, for H and a continuous group of unitary operators in H, $\{\tau(t), \ t \ \epsilon \ \mathbb{R}\}$ (i.e., an unitary representation of the group $\mathbb{R}$), $[H; \ \tau(t), \ t \ \epsilon \ \mathbb{R}]$ is called a _continuous evolution system._

Given a pair $[H_1; \ \tau_1(t), \ t \ \epsilon \ \mathbb{R}]$ and $[H_2; \ \tau_2(t), \ t \ \epsilon \ \mathbb{R}]$ of continuous evolution systems, a sesquilinear form $B: H_1 \times H_2 \rightarrow \mathbb{C}$ is _τ-Toeplitz_ if for all $(f_1, f_2) \ \epsilon \ H_1 \times H_2$,

$$B(\tau_1(t)f_1, f_2) = B(f_1, \tau_2(-t)f_2), \qquad \forall t \geq 0 \qquad (1.1)$$

Fixing a pair of closed subspaces $W_1 \subset H_1$ and $W_2 \subset H_2$, satisfying

$$\tau_1(t)W_1 \subset W_1 \text{ and } \tau_2(-t)W_2 \subset W_2, \ \forall t \geq 0 \qquad (1.2)$$

a sesquilinear form $B: W_1 \times W_2 \rightarrow \mathbb{C}$ is _τ-Hankel_ if (1.1) is satisfied only for all $(f_1, f_2) \ \epsilon \ W_1 \times W_2$.

Similarly, in the discrete case, $B: H_1 \times H_2 \rightarrow \mathbb{C}$ is _τ-Toeplitz_ if for all $(f_1, f_2) \ \epsilon \ H_1 \times H_2$,

$$B(\tau_1^n f_1, f_2) = B(f_1, \ \tau_2 f_2), \qquad \forall n \geq 0 \qquad (1.1a)$$

and, if

$$\tau_1^n W_2 \subset W_2, \quad \tau_1^{-n} W_2 \subset W_2, \qquad \forall n \geq 0 \qquad (1.2a)$$

$B: W_1 \times W_2 \rightarrow \mathbb{C}$ is _τ-Hankel_ if it satisfies (1.1a) only for all $(f_1, f_2) \ \epsilon \ W_1 \times W_2$.

The pairs of systems consisting of two copies of the same evolution system are of special interest. In particular the basic example is the <u>trigonometric case</u> where $H_1 = H_2 = L^2(\mathbb{R})$, $\tau_1(t) = \tau_2(t) = \tau(t)$, $\tau(t)f(x) = e^{itx}f(x)$ is the (multiplication) <u>shift</u> of $L^2(\mathbb{R})$. Here the subspaces $W_1 = H^2(\mathbb{R})$, $W_2 = H^2_-(\mathbb{R}) = H^2(\mathbb{R})^\perp$ satisfy (1.2), and the τ-Hankel forms B: $W_1 \times W_2 \to \mathbb{C}$ with domain $H^2(\mathbb{R}) \times H^2_-(\mathbb{R})$ coincides with the classical Hankel forms. However, <u>even in this trigonometric case, our class of Hankel forms is considerably more general than the classical one</u>. This is so since the pair W_1, W_2 may vary. Observe that in the trigonometric case, the class of τ-Toeplitz Toeplitz forms is the same as the classical one.

Hankel forms in discrete evolution systems where studied in previous papers (see [CS1], [CS4]). Our next task is to show that the well-known characterizations of the generators and cogenerators of continuous semigroups [Da] allow to reduce the consideration of Hankel forms in continuous evolution systems to the corresponding one in discrete systems, as follows.

Let us recall that a linear operator γ in a Hilbert space H is an <u>isometry</u> if its domain $D(\gamma)$ and its range $R(\gamma)$ are both closed subspaces of H, and $\|\gamma(x)\| = \|x\|$, $\forall x \in D(\gamma)$. If in addition $D(\gamma) = H$, γ is called <u>semiunitary</u>. The Cayley transform $\gamma = (A - i)(A + i)^{-1}$ establishes a bijection $\gamma \longmapsto A$ between the closed symmetric operators A with domain $D(A)$ dense in H and the isometries γ such that $(\gamma - 1)(H)$ is dense in H. Moreover, if $\gamma \longmapsto A$, the following condition are equivalent:

(a) γ is unitary;

(b)   A is self-adjoint, so that $(A \pm i\lambda)D(A) = H$ and $(A \pm i\lambda)^{-1}$: $H \to D(A)$ is bounded $\forall \lambda > 0$;

(c)   A is the <u>generator</u> of a semigroup $\{\tau(t); t \geq 0\}$ of unitary operators, i.e., $Af = \frac{1}{i}\frac{d}{dt}\tau(t)f\big|_{t=0}$, $\forall f \in D(A)$, and in this case γ is called the <u>cogenerator</u> of $\{\tau(t)\}$.

Thus γ is a cogenerator of a unitary group iff γ is unitary and $(\gamma - 1)(H)$ is dense in H.

Furthermore (see for instance [Da]), the following conditions are equivalent:

(a$_1$) γ is semiunitary;

(b_1) A is symmetric and $(A + i\lambda)D(A) = H$, so that

$(A + i\lambda)^{-1}: H \to D(A)$ is bounded, $\forall \lambda > 0$;

(c_1) A is the generator of a semigroup $\{\tau(t); t \geq 0\}$ of

semiunitary operators in H.

Thus γ is a cogenerator of a semiunitary semigroup iff γ is semiunitary and $(\gamma - 1)(H)$ is dense in H.

Finally, if A is the generator of the unitary group $\{\tau(t): t \in \mathbb{R}\}$, then, $\forall f \in H$, $\lambda > 0$,

$$(A + i\lambda)^{-1}f = \int_0^\infty e^{-\lambda t}\tau(t)f(t)dt,$$

and   (1.3)

$$(A - i\lambda)^{-1}f = \int_0^\infty e^{-\lambda t}\tau(-t)f(t)dt$$

PROPOSITION 1.1. *Let $\{\tau(t): t \in \mathbb{R}\}$ be a group of unitary operators in H, A the generator and τ the cogenerator of the group. Then for every closed subspace $W \subset H$ the following conditions are equivalent:*

(i) $\tau(t)W \subset W$, $\forall t \geq 0$;

(ii) $\tau W \subset W$ *and* $(\tau - 1)W$ *is dense in* W.

If these conditions are satisfied, then $D(A) \cap W$ is dense in W, and $D(A) \cap W = D(A^\circ)$, where A° is the generator of the semiunitary semigroup $\{\tau^\circ(t) = \tau(t)|_W: t \geq 0\}$ in W, and $A = A^\circ$ in $D(A^\circ)$.

PROOF. (i) <u>implies</u> (ii). Setting, for $t \geq 0$, $\tau^\circ(t) = \tau(t)|_W$, $\{\tau^\circ(t): t \geq 0\}$ is a semiunitary semigroup acting in W, whose generator A° has domain $D(A^\circ) = D(A) \cap W$ and coincides with A in $D(A^\circ)$, while its

cogenerator τ° is semiunitary and satisfies $(\tau^\circ - 1)W = W$, so that the last assertion of the Proposition holds. To prove assertion (ii), let us see that $\tau = \tau^\circ$ in W.

For each $f \in W$ there exist $g \in D(A)$ and $g^\circ \in D(A^\circ)$ such that $f = (A + i)g = (A^\circ + i)g^\circ = (A + i)g^\circ$, so that $g = g^\circ = (A + i)^{-1}f$, hence $\tau f = (A - i)g = (A^\circ - i)g = (A^\circ - i)g^\circ = \tau^\circ f$, and $\tau f = \tau^\circ f$, as claimed. (ii) <u>implies</u> (i). Setting $\tau^\circ = \tau|_W$, τ° is semiunitary in W, and $(\tau^\circ - 1)W$ is dense in W, so by the equivalence $(a_1) \leftrightarrow (b_1)$ above, τ° is the Cayley transform of a symmetric operator A° in W, $\tau^\circ = (A^\circ - i)(A^\circ + i)^{-1}$, and

$(A^\circ + i\lambda)D(A) = W$ for $\lambda > 0$, $D(A^\circ) = (\tau^\circ - 1)W$.

Therefore, every $g \in D(A^\circ)$ is of the form

$g = (\tau^\circ - 1)f = (\tau - 1)f$ for some $f \in W$, and $A^\circ g = (\tau^\circ + 1)(\tau^\circ - 1)^{-1}g$

$= (\tau + 1)f = (\tau + 1)f = (\tau + 1)(\tau - 1)^{-1}g = Ag$. Thus, $Ag = A^\circ g$ for all

$g \in D(A^\circ)$.

If $\lambda > 0$ and $g \in D(A^\circ)$, then $(A + \lambda)^{-1}(A^\circ + \lambda)g$

$= (A + \lambda)^{-1}(A + \lambda)g = g$, i.e., $(A + \lambda)^{-1}f = g \in W$ for all $f \in W$. Hence,

$\langle(A + \lambda)^{-1}f, f'\rangle = 0$ for all $f \in W$, $f' \in W^\perp$, $\lambda > 0$, and, by (1.3), the

Laplace transform of $\langle\tau(t)f, f'\rangle$ is zero,

$$\int_0^\infty e^{-\lambda t} \langle\tau(t)f, f'\rangle dt = 0 \text{ for } \lambda > 0.$$

By the uniqueness of the Laplace transform, $\langle\tau(t)f, f'\rangle = 0$ for all $f \in W$,

$f' \in W^\perp$, $t \geq 0$, and, thus, $\tau(t)f \in W$ for all $t \geq 0$, as claimed. $\square$

To each pair $[H_1; \tau_1(t), t \in \mathbb{R}]$ and $[H_2; \tau_2(t), t \in \mathbb{R}]$ of

continuous evolution systems we associate a discrete pair $[H_1; \tau_1^n, n \in \mathbb{Z}]$

and $[H_2; \tau_2^n, n \in \mathbb{Z}]$, where τ_1 is the cogenerator of the unitary semigroup

$\{\tau_1(t), t \geq 0\}$, and τ_2^{-1} is the cogenerator of the semigroup

$\{\tau_2(-t): t \geq 0\}$.

LEMMA 1.2. *Let* $[H_1; \tau_1(t), t \in \mathbb{R}]$, $[H_2; \tau_2(t), t \in \mathbb{R}]$ *be a pair of*

continuous evolution systems, and $[H_1, \tau_1^n]$, $[H_2, \tau_2^n]$ *be its associated*

discrete pair, and let A_1, A_2 *be the generators of* $\{\tau_1(t): t \in \mathbb{R}\}$,

$\{\tau_2(t): t \in \mathbb{R}\}$, *respectively, so* $A_i \longmapsto \tau_i$, $i = 1,2$. *Then, for every*

bounded sesquilinear form $B: W_1 \times W_2 \to \mathbb{C}$, $W_i \subset H_i$, *closed subspaces,*

$i = 1,2$, *the following conditions are equivalent:*

   (1) B *is* $\tau(t)$-*Hankel*

   (2) *If* $t \geq 0$, $\lambda > 0$, *then* $\tau_1(t)W_1 \subset W_1$, $\tau_2(-t)W_2 \subset W_2$, $(A_1 + \lambda i)$

   $(D(A_1) \cap W_1) = W_1$, $(A_2 - \lambda i)(D(A_2) \cap W_2) = W_2$, *and*

   $B(A_1 g_1, g_2) = B(g_1, A_2 g_2)$, *for all*

   $g_1 \in D(A_1) \cap W_1$, $g_2 \in D(A_2) \cap W_2$.   (1.4)

(3) *If* $t \geq 0$, $\lambda > 0$, *then* $\tau_1(t)W_1 \subset W_1$, $\tau_2(-t)W_2 \subset W_2$,

$(A_1 + \lambda i)^{-1}W_1 = D(A_1) \cap W_1$, $(A_2 - \lambda i)^{-1}W_2 = D(A_2) \cap W_2$, *and*

$B((A_1 + \lambda i)^{-1}f_1, f_2) = B(f_1, (A_2 - \lambda i)^{-1}f_2)$, *for all*

$(f_1, f_2) \in W_1 \times W_2$ (1.5)

$\overline{(4)}$ B *is* τ-*Hankel and* $\overline{(\tau_1 - 1)W_1} = W_1$, $\overline{(\tau_2^{-1} - 1)W_2} = W_2$.

PROOF. (1) <u>implies</u> (2). Since (1) implies in particular $\tau_1(t)W_1 \subset W_1$ and
$\tau_2(-t)W_2 \subset W_2$ for $t \geq 0$, (2) follows immediately from (1), Proposition 1.1
and the definition

$$A_j g_j = \lim_{t \to 0} \frac{\tau_j(t)g_j - g_j}{it}, \ j = 1,2.$$

(2) <u>implies</u> (3). Since by (1.4), for all $\lambda > 0$,
$(f_1, f_2) \in W_1 \times W_2$, $B((A_1 + \lambda i)^{-1}f_1, f_2) = B((A_1 + \lambda i)^{-1}f_1,$
$(A_2 - \lambda i)(A_2 - \lambda i)^{-1}f_2) = B((A_1 + \lambda i)(A_1 + \lambda i)^{-1}f_1,$
$(A_2 - \lambda i)^{-1}f_2) = B(f_1, (A_2 - \lambda i)^{-1}f_2)$, (1.5) holds.

(3) <u>is equivalent to</u> (4). This follows from Proposition 1.1,
(1.5) and $\tau_1 = 1 - 2i(A_1 + \lambda i)^{-1}$, $\tau_2^{-1} = 1 - 2i(A_2 - \lambda i)^{-1}$,
since B is (τ_1, τ_2)-Hankel means $\tau_1 W_1 \subset W_1$, $\tau_2^{-1} W_2 \subset W_2$ and
$B(\tau_1 f_1, f_2) = B(f_1, \tau_2^{-1} f_2)$ for $(f_1, f_2) \in W_1 \times W_2$.

(3) <u>implies</u> (1). Let $f_1 \in W_1$, $f_2 \in W_2$ be fixed. Setting
$F(t) = B(\tau_1(t)f_1, f_2)$, $G(t) = B(f_1, \tau_2(-t)f_2)$, by (1.3), the
assertion (1.5) of (3) can be written as
$$L(F)(\lambda) = L(G)(\lambda) \text{ for } \lambda > 0$$
where L is the Laplace transform. By the uniqueness of the
Laplace transform, $F(t) = G(t)$, i.e., $B(\tau_1(t)f_1, f_2)$
$= B(f_1, \tau_2(-t)f_2)$ for $(f_1, f_2) \in W_1 \times W_2$. □

COROLLARY 1.3. *The study of Hankel and Toeplitz forms in a pair of*
continuous evolution systems reduces to that of the same Hankel and
Toeplitz forms in its associated discrete pair.

2. THE A-A-K THEOREM IN GENERAL AND REGULAR REPRESENTATIONS OF R.

Let $[H_1; \tau_1(t), t \in \mathbb{R}]$, $[H_2; \tau_2(t), t \in \mathbb{C}]$ be a pair of continuous evolution systems, $B: W_1 \times W_2 \to \mathbb{C}$, a bounded Hankel form with respect to the pair (see Section 1 for its definition), and $\Gamma: W_1 \to W_2$ its associated operator, $\Gamma \sim B$, defined by

$$\langle \Gamma f_1, f_2 \rangle = B(f_1, f_2), \qquad \forall (f_1, f_2) \in W_1 \times W_2. \qquad (2.1)$$

For each nonnegative integer n, let the n-singular numbers of B be defined by

$$s_n(B) := s_n(\Gamma) = \inf \{\|\Gamma - T\|: \text{rank } T \leq n\}, \qquad (2.2)$$

so that $s_0(B) = \|B\|$.

A Toeplitz form $B': H_1 \times H_2 \to \mathbb{C}$ is called a lifting of B if $\|B'\| = \|B\| = s_0(B)$ and $B' = B$ in $W_1 \times W_2$. A Toeplitz form $B^{(n)}$ is called an n-conditional lifting of B if $\|B^{(n)}\| = s_n(B)$ and there exists a subspace M of W_1, codim $M \leq n$, such that $B^{(n)} = B$ in $M \times W_2$. Thus, for n = 0, a 0-conditional lifting $B^{(0)} = B'$ is a lifting.

THEOREM 2.1. *Given a pair of continuous one-parameter evolution systems* $[H_1; \tau_1(t), t \in \mathbb{R}]$ *and* $[H_2; \tau_2(t), t \in \mathbb{R}]$, *and a bounded sesquilinear Hankel form* $B: W_1 \times W_2 \to \mathbb{C}$, *there exist a Toeplitz lifting B' of B, as well as an n-conditional lifting* $B^{(n)}$ *of B, for all* $n \geq 0$. *Moreover,*

$$s_n(B) := s_n(\Gamma) = \inf\{\|\Gamma - \Gamma_n\|: \Gamma_n \text{ Hankel, rank } \Gamma_n \leq n\} \quad (2.3)$$

where Γ is the operator associated to B by (2.1), and the infima is attained.

The discrete analogue of Theorem 2.1 was proved in [CS4, Thm. 1 & Cor. 1] and therefore the theorem is an immediate consequence of Corollary 1.3.

Formula (2.3) gives an abstract version, for continuous evolution systems, of the classical A-A-K theorem in the line.

The statement of Theorem 2.1 can be made more precise when the pair of evolution systems consists of two copies of the regular

representation of $\mathbb{R}$, i.e., in the case the pair is $[H_i; \tau_i(t), t \in \mathbb{R}]$,
$i = 1,2$, where $H_1 = H_2 = L^2(\mathbb{R})$ and $\tau_1 = \tau_2 = \tau$ is the shift operator,
$\tau(t)f(x) = e^{itx}f(x)$. Let $W_1, W_2 \subset L^2(\mathbb{R})$ and $B: W_1 \times W_2 \to \mathbb{C}$ be a bounded
Hankel form, so that $\tau(t)W_1 \subset W_1$, $\tau(-t)W_2 \subset W_2$,
$B(\tau(t)f_1, f_2) = B(f_1, \tau(-t)f_2)$, $\forall t \geq 0$. The basic case is when $W_1 = H^2(\mathbb{R})$,
$W_2 = H_-^2(\mathbb{R}) = H^2(\mathbb{R})^\perp$, and in this case a Hankel form B is just the classical
one and $\Gamma \sim B$ as in (2.1) is a Hankel operator. By Theorem 2.1 there
exists a Toeplitz form $B': L^2(\mathbb{R}) \times L^2(\mathbb{R}) \to \mathbb{C}$ such that $B' = B$ in
$H^2(\mathbb{R}) \times H_-^2(\mathbb{R})$, and $\|B'\| = \|B\|$. Moreover, if Γ' is the operator associated
to B', then the Toeplitz condition on B', i.e., $B'(\tau(t)f_1, f_2)$
$= B'(f_1, \tau(-t)f_2)$ for all t and all $f_1, f_2 \in L^2(\mathbb{R})$, means that Γ' commutes
with the shift $\tau(t)$ and, therefore, as well-known (cf. [H]),

$$B'(f_1, f_2) = \int f_1 \bar{f}_2 \phi \, dx, \text{ for } \phi \in L^\infty, \|\phi\|_\infty = \|B'\| = \|B\|, \tag{2.4}$$

so that B has the integral representation

$$B(f_1, f_2) = \int f_1 \bar{f}_2 \phi \, dx, \ (f_1, f_2) \in H^2(\mathbb{R}) \times H_-^2(\mathbb{R}) \tag{2.5}$$

The bounded functions ϕ for which such integral representations hold are
called symbols of the Hankel form B, or of the associated Hankel operator
Γ. Thus, for every symbol ϕ of B, $\|B\| \leq \|\phi\|_\infty$, and there is a symbol ψ
such that $\|B\| = \|\psi\|_\infty$.

By the well-known theorems of Beurling and Lax (cf. [H]), the
study of the invariant subspaces $L \subset L^2(\mathbb{R})$, $\tau(t)L \subsetneq L$ for $t \geq 0$ can be
reduced to that of the invariant subspaces $M \subset L^2(T)$, and $M = qH^2(T)$,
$|q| = 1$, while $L = \theta H^2(\mathbb{R})$, $|\theta| = 1$. If in addition $M \subset H^2(T)$, then $q \in H^\infty$
and, if, M has codimension n in $H^2(T)$, then θ is a product of n Blaschke
factors $|\lambda|(\lambda - e^{ix})/\lambda(1 - \bar{\lambda}e^{ix})$, $|\lambda| < 1$. If $\Gamma_n: H^2(\mathbb{R}) \to H_-^2(\mathbb{R})$ is a
Hankel operator of rank $\leq$ n, then $L = \ker \Gamma_n \subset H^2(\mathbb{R})$ is invariant under
$\tau(t)$, and has codimension $\leq$ n. Therefore $L = bH^2(\mathbb{R})$, where b is a product
of n Blaschke factors on $\mathbb{R}$, $(x - \mu)(x - \bar{\mu})^{-1} |1 + \mu^2| (1 + \mu^2)^{-1}$, Im $\mu \geq 0$,
$\Gamma(bf) = 0$ for all $f \in H^2(\mathbb{R})$, and, if ϕ_n is a symbol of $B_n \sim \Gamma_n$, then

$B_n(bf_1, f_2) = \int f_1 \bar{f}_2 \, b\phi_n dx = 0$ for all $f_1 \in H^2(\mathbb{R})$, $f_2 \in H^2_-(\mathbb{R})$, which gives

$b\phi_n = h \in H^\infty$. Thus, the Hankel forms of rank $\leq n$ are given by symbols of

the form $\phi_n = \overset{*}{b} h$, $h \in H^\infty$. Formula (2.3) of Theorem 2.1 then gives the

classical A-A-K theorem:

$$s_n(B) = \inf\{\|\phi - \phi_n\|_\infty : \phi_n = \overset{*}{b} h \in H^\infty_1, b \text{ n-Blaschke product}\}$$

where ϕ is a symbol of B.

If W_1, W_2 are <u>arbitrary</u> subspaces of $L^2(\mathbb{R})$, satisfying

$\tau(t)W_1 \subset W_1$, $\tau(-t)W_2 \subset W_2$, $\forall t \geq 0$, with strict inclusions, then, by the

Beurling-Lax theorem, $W_1 = \theta_1 H^2(\mathbb{R})$, $W_2 = \theta_2 H^2(\mathbb{R})$, and this general case is

easily reduced to the previous one, giving more precise statements for

Theorem 2.1.

REMARK 2.1. If $\{U(t): t \in \mathbb{R}\}$ is a unitary group in a Hilbert space N, and

if $N_1, N_2 \subset N$ are such that $U(t)N_1 \subset N_1$, $U(-t)N_2 \subset N_2$, for all $t \geq 0$, then

$\{U(t)\}$ is called a <u>unitary coupling</u> of the two semiunitary semigroups

$\{U(t)\big|_{N_1} : t \geq 0\}$ and $\{U(-t)\big|_{N_2} : t \geq 0\}$. The first assertion of Theorem

2.1, about the existence of a Toeplitz lifting B' for the Hankel form B,

can be expressed in terms of unitary couplings as follows: Given a pair of

evolution systems, $[H_1; \tau_1(t), t \in \mathbb{R}]$ and $[H_2; \tau_2(t), t \in \mathbb{R}]$, a pair of

closed subspace $W_1 \subset H_1$, $W_2 \subset H_2$ such that $\tau_1(t)W_1 \subset W_1$, $\tau_2(-t)W_2 \subset W_2$, for

all $t \geq 0$, and a unitary coupling $U(t) \in L(W)$, $W \supset W_1, W_2$, of the semigroups

$\{\tau_1(t)\big|_{W_1} : t \geq 0\}$, $\{\tau_2(-t)\big|_{W_2} : t \geq 0\}$, then there exists a unitary coupling

$T(t) \in L(H)$, $H \supset H_1, H_2$, of the groups $\{\tau_1(t): t \in \mathbb{R}\}$, $\{\tau_2(t): t \in \mathbb{R}\}$ such

that $\langle f_1, f_2 \rangle_H = \langle f_1, f_2 \rangle_W$ for all $(f_1, f_2) \in W_1 \times W_2$. (See [CS3] for the

discrete case.)

3. THE A-A-K THEOREM FOR GENERAL REPRESENTATIONS OF $\mathbb{R}^2$ AND $(\mathbb{R}^2, [,])$.

We say that $[H; \sigma(t), \tau(t), t \in \mathbb{R}]$ is a continuous two-parameter

evolution system if $\{\sigma(t), t \in \mathbb{R}\}$ and $\{\tau(t), t \in \mathbb{R}\}$ are two continuous

groups of unitary operators in the Hilbert space H. In this context, the

evolution system of the previous sections will be referred as one-parameter evolution systems.

We assume

$$\sigma(t)\tau(t) = e^{i\alpha}\tau(t)\sigma(t), \quad \forall t \in \mathbb{R} \text{ and some } \alpha \in \mathbb{R} \quad (3.1)$$

If $\alpha = 0$, $[H;\sigma(t),\tau(t),t \in \mathbb{R}]$ is a unitary representation of $\mathbb{R}^2$. If $\alpha = \pi$, $[H;\sigma(t),\tau(t),t \in \mathbb{R}]$ is a unitary representation of $(\mathbb{R}^2, [\,,\,])$.

Given a pair of two-parameter evolution systems, $[H_1; \sigma_1(t), \tau_1(t), t \in \mathbb{R}]$ and $[H_2; \sigma_2(t), \tau_2(t), t \in \mathbb{R}]$, a sesquilinear form $B: H_1 \times H_2 \longrightarrow \mathbb{C}$ is called <u>Toeplitz</u> if it is both σ-Toeplitz and τ-Toeplitz, in the sense of definition (1.1).

If $W_1 \subset H_1$, $W_2 \subset H_2$ are closed subspaces satisfying, for all $t \geq 0$,

$$\sigma_1(t)W_1 \subset W_1 \quad \text{and} \quad \sigma_2^{-1}(t)W_2 \subset W_2 \quad (1.2b)$$

$$\tau_1(t)W_1 \subset W_1 \quad \text{and} \quad \tau_2^{-1}(t)W_2 \subset W_2 \quad (1.2)$$

a form $B: W_1 \times W_2 \longrightarrow \mathbb{C}$ is called <u>Hankel</u> if it is both σ-Hankel and τ-Hankel.

Of special interest among the two-parameter evolution systems are (1) the <u>two-parameter</u> <u>trigonometric case</u>, where $H_1 = H_2 = L^2(\mathbb{T}^2)$ and $\sigma_1 = \sigma_2 = \sigma$, $\tau_1 = \tau_2 = \tau$ are the two shifts in $\mathbb{T}^2$, $\sigma f(x,y) = e^{ix}f(x,y)$ and $\tau f(x,y) = e^{iy}f(x,y)$; (2) the <u>plane case</u>, where $H_1 = H_2 = L^2(\mathbb{R}^2)$, $\sigma_1 = \sigma_2 = \sigma$, $\tau_1 = \tau_2 = \tau$, and $\sigma(t)f(x,y) = e^{itx}f(x,y)$, $\tau(t)f(x,y) = e^{ity}f(x,y)$, $t \in \mathbb{R}$, and, 3) the <u>symplectic case</u>, where $H_1 = H_2 = L^2(\mathbb{R}^2) \cong L^2(\mathbb{C})$, $\sigma_1 = \sigma_2 = \sigma$, $\tau_1 = \tau_2 = \tau$ are the <u>twisted</u> shifts $\sigma(t)f(x,y) = e^{-itx}f(x + t,y)$, $\tau(t)f(x,y) = e^{itx}f(x,y + t)$, $t \in \mathbb{R}$.

Let now $[H_1;\sigma_1(t),\tau_1(t),t \in \mathbb{R}]$, $[H_2;\sigma_2(t),t \in \mathbb{R}]$ be a pair of continuous two-parameters systems, and $B: W_1 \times W_2 \rightarrow \mathbb{C}$ a bounded sesquilinear Hankel form in it. In this situation, $\sigma_2(-t)W_2 \subset W_2$ for all $t \geq 0$, but not necessarily for <u>all</u> t, and the subspaces

$$W_2^\sigma = \{f \in W_2: \sigma_2(t)f \in W_2, \forall t \in \mathbb{R}\}$$

and $\hspace{8cm} (3.2)$

$$W_2^\tau = \{f \in W_2: \tau_2(t)f \in W_2, \forall t \in \mathbb{R}\}$$

will be of importance. Indeed, here <u>we consider only</u> Hankel forms

B: $W_1 \times W_2 \to C$ for which

$$W_2 = \{f + g: f \in W_2^\sigma, g \in W_2^\tau\} \qquad (3.2a)$$

and sometimes will require also

$$W_2 = \{f + g: f \in W_2^\sigma, g \in W_2^\tau, f \perp g\}. \qquad (3.2b)$$

Given a bounded Hankel B: $W_1 \times W_2 \to C$, a pair of Toeplitz forms

B′, B″ is a <u>lifting pair</u> for B if

$$\|B'\| \leq \|B\|, \quad \|B''\| \leq \|B\|,$$

and   (3.3)

$$B' = B \text{ in } W_1 \times W_2^\sigma, \qquad B'' = B \text{ in } W_1 \times W_2^\tau.$$

By (3.2a), the form B is determined by any of its lifting pairs B′, B″.

For each bounded form B′: $H_1 \times H_2 \longrightarrow C$, since the function

$F(t) = B'(\sigma_1(t)f, \sigma_2(t)g)$ is bounded for each pair $(f,g) \in H_1 \times H_2$,

$|F(t)| \leq \|B\| \|f\|_{H_1} \|g\|_{H_2}$, a form B^σ: $H_1 \times H_2 \longrightarrow C$ can be defined in $H_1 \times H_2$

by

$$B^\sigma(f,g) := \text{LIM}_{t \to \infty} F(t) = \text{LIM}_{t \to \infty} B'(\sigma_1(t)f, \sigma_2(t)g), \qquad (3.4)$$

where LIM stands for the Banach-Mazur limit as $t \to \infty$. Then $\|B^\sigma\| \leq \|B'\|$,

and, by the properties of LIM, it follows easily that B^σ is always

σ-Toeplitz, and that, if B′ is itself τ-Toeplitz, then B^σ is both σ- and

τ-Toeplitz hence, Toeplitz (cfr. [DS] and [CS4]).

Similarly, if B: $W_1 \times W_2 \longrightarrow C$ is bounded, where W_1 and W_2 satisfy

(1.2), given T: $W_1 \longrightarrow W_2$, the operator associated to B by (2.1), then an

operator T^σ: $W_1 \longrightarrow W_2^\sigma$ can be defined by

$$\langle T^\sigma f, g \rangle := \text{LIM}_{t \to \infty} \langle T\sigma_1(t)f, \sigma_2(t)g \rangle \qquad (3.4a)$$

and $\|T^\sigma\| \leq \|T\|$. Moreover, the form B^σ, associated to T^σ, is always

σ-Hankel and, if B is itself τ-Hankel in $W_1 \times W_2$, then B^σ is both σ- and

τ-Hankel in $W_1 \times W_2^\sigma$ (cf. [CS4; Lemma 1]).

The operator T^τ: $W_1 \to W_2^\tau$, and its associated form

B^τ: $W_1 \times W_2^\tau \to C$, are similarly defined. Set

$$F_n = F_n(W_1, W_2) = \{T: W_1 \to W_2, \text{ linear op. of rank } \leq n\} \qquad (3.5)$$

$$F_{n\tau} = \{T \in F_n: T \text{ is } \tau\text{-Hankel}\} \tag{3.5a}$$

$$F_{n\tau}^{\sigma} = \{T^{\sigma}: T \in F_{n\tau}\} \tag{3.5b}$$

and, similarly, $F_{n\sigma}$ and $F_{n\sigma}^{\tau}$. Note that all operators in $F_{n\tau}^{\sigma}$ and $F_{n\sigma}^{\tau}$ are Hankel, but not necessarily of finite rank.

THEOREM 3.2. *Given a pair of continuous two-parameter evolution systems* $[H_1; \sigma_1(t), \tau_1(t), t \in \mathbb{R}]$, $[H_2; \sigma_2(t), \tau_2(t), t \in \mathbb{R}]$, *and a bounded sesquilinear Hankel form* $B: W_1 \times W_2 \to \mathbb{C}$, *there exists a lifting pair* B', B'' *for* B. *Moreover, if* Γ *is the operator associated to* B *by (2.1), for every integer* $n \geq 0$,

$$\inf \{\|P_{\sigma}\Gamma - \Gamma_n'\|: \Gamma_n' \in F_{n\tau}^{\sigma}\} \leq s_n(\Gamma) \tag{3.6}$$

$$\inf \{\|P_{\tau}\Gamma - \Gamma_n''\|: \Gamma_n'' \in F_{n\sigma}^{\tau}\} \leq s_n(\Gamma) \tag{3.6a}$$

and

$$s_n(\Gamma) = \inf \{\|\Gamma - \Gamma_n\|: \Gamma_n \in F_{n\tau}\} = \inf \{\|\Gamma - \Gamma_n\|: \Gamma_n \in F_{n\sigma}\} \tag{3.6b}$$

where P_{σ}, P_{τ} *are the orthoprojectors of* W_2 *onto* W_2^{σ} *and* W_2^{τ}, *respectively, and the infima are attained.*

PROOF. Since B is Hankel, it is in particular τ-Hankel, thus, by Theorem 2.1 for n = 0, there exists a τ-Toeplitz form $B_1': H_1 \to \mathbb{C}$, such that $B_1' = B$ in $W_1 \times W_2$, with $\|B_1'\| = \|B\|$. Setting $B' = (B_1')^{\sigma}$ (see (3.4)), B' is both σ- and τ-Toeplitz, hence Toeplitz. Since by the definition (3.2) of W_2^{σ}, for $(f,g) \in W_1 \times W_2^{\sigma}$, $\forall t \geq 0$,

$$B(f,g) = B(f, \sigma_2(-t)\sigma_2(t)g) = B(\sigma_1(t)f, \sigma_2(t)g),$$

it follows from (3.4) that $B' = B$ in $W_1 \times W_2^{\sigma}$. Similarly, B'' can be defined as $(B_1'')^{\tau}$, where B_1'' is the σ-Toeplitz lifting of B, considered a σ-Hankel form. Thus, B' and B'' form a lifting pair for B. The equalities (3.6b) follow from (2.3) in Theorem 2.1 by ignoring either $\{\sigma_1(t): t \in \mathbb{R}\}$ and $\{\sigma_2(t): t \in \mathbb{R}\}$ or $\{\tau_1(t): t \in \mathbb{R}\}$ and $\{\tau_2(t): t \in \mathbb{R}\}$, and considering the given systems as one-parameter ones. The inequalities (3.6) and (3.6a) follow by the same argument as in [CS4, Theorem 4], using the properties of the Banach-Mazur limits that define the operators in $F^T_{n\sigma}$ and $F^\sigma_{n\tau}$. □

For W_1 and W_2 as above, we say that W^σ_2 satisfies the weak Riemann-Lebesgue condition if for every pair $f,g \in W^\sigma_2$ the following ordinary limit exists:
$$\lim_{t\to\infty} \langle f, \sigma_2(t)g\rangle = f^\sigma(g). \tag{3.7}$$
For each fixed f, f^σ is a bounded antilinear functional in W^σ_2, thus, $\forall f \in W^\sigma_2$, $\exists f^\sigma \in W^\sigma_2$ such that
$$f^\sigma(g) = \langle f^\sigma, g\rangle, \quad \forall g \in W^\sigma_2. \tag{3.8}$$
If (3.7) holds with $f^\sigma = 0$, $\forall f \in W^\sigma_2$, then W^σ_2 is said to satisfy the Riemann-Lebesgue condition. Similar conditions are defined for W^T_2. By the same argument as in the discrete case (see [CS4]) follows

LEMMA 3.2. *If W^σ_2 satisfies the weak Riemann-Lebesgue condition (3.7), then every $T^\sigma \in F^\sigma_{n\tau}$ has finite rank $\leq n$. Moreover, if W^σ_2 satisfies the Riemann-Lebesgue condition, then all operators $T^\sigma \in F^\sigma_{n\tau}$ have rank zero, i.e., $T^\sigma \equiv 0$. The same results hold for W^T_2 and the operators in $F^T_{n\sigma}$.*

From Theorem 3.1 and Lemma 3.2 follows

THEOREM 3.3. *Let $[H_1; \sigma_1(t), \tau_1(t), t \in \mathbb{R}]$, $[H_2; \sigma_2(t), \tau_2(t), t \in \mathbb{R}]$ be a pair of continuous systems, and $W_1 \subset H_1$, $W_2 \subset H_2$, be closed subspaces such that (1.2) and (1.2b) hold and W^σ_2 and W^T_2 satisfy the weak Riemann-Lebesgue condition. Then every Hankel form $B: W_1 \times W_2 \to \mathbb{C}$ has an associated operator $\Gamma: W_1 \to W_2$ for which*
$$\inf\{\|P_\sigma\Gamma - \Gamma'\|: \Gamma' \text{ Hankel operator of rank} \leq n\} = s_n(P_\sigma\Gamma) \leq s_n(\Gamma)$$
and
$$\tag{3.9}$$
$$\inf\{\|P_\tau\Gamma - \Gamma''\|: \Gamma'' \text{ Hankel operator of rank} \leq n\} = s_n(P_\tau\Gamma) \leq s_n(\Gamma)$$

where P_σ and P_τ are the orthoprojectors of W_2 onto W_2^σ and W_2^τ, respectively.

In particular, if W_2^σ and W_2^τ satisfy the Riemann Lebesgue condition, then, for all $n \geq 0$,

$$\|P_\sigma \Gamma\| \equiv s_0(P_\sigma \Gamma) = s_n(P_\sigma \Gamma) \leq s_n(\Gamma)$$

and   (3.9a)

$$\|P_\tau \Gamma\| \equiv s_0(P_\tau \Gamma) = s_n(P_\tau \Gamma) \leq s_n(\Gamma).$$

In this case, if Γ is also compact, then $\Gamma = 0$, i.e., there are no nonzero compact Hankel operators $\Gamma \colon W_1 \to W_2$ with W_2 satisfying (3.2a), and W_2^σ and W_2^τ satisfying the Riemann-Lebesgue condition.

4. A-A-K THEOREMS IN THE REGULAR REPRESENTATION OF $\mathbb{R}^2$.

Theorems 3.2 and 3.3 can be given more precise statements when the pair of evolution systems consists of two copies of $[H; \sigma(t), \tau(t), t \in \mathbb{R}]$, the <u>regular representation of $\mathbb{R}^2$</u>, that is where $H = L^2(\mathbb{R}^2)$, and σ and τ are the shifts defined by

$$\sigma(t)f(x,y) = e^{itx}f(x,y), \quad \tau(t)f(x,y) = e^{ity}f(x,y). \quad (4.1)$$

Let $B \colon W_1 \times W_2 \to \mathbb{C}$ be a bounded Hankel form, where W_1 and W_2 must verify that, for all $t \geq 0$, W_1 is both $\tau(t)$ and $\sigma(t)$ invariant, W_2 is both $\tau(-t)$ and $\sigma(-t)$ invariant, and $W_2 = W_2^\sigma \oplus W_2^\tau$, as in (3.2b).

By Theorem 3.2, there exists a pair of Toeplitz forms $B', B'' \colon L^2(\mathbb{R}^2) \times L^2(\mathbb{R}^2) \to \mathbb{C}$, such that $B' = B$ in $W_1 \times W_2^\sigma$, $B'' = B$ in $W_1 \times W_2^\tau$ and $\|B'\|, \|B''\| \leq \|B\|$. Assigning to each $f \in L^2(\mathbb{R}^2)$ a function $F \colon x \to f_y(x) = f(x,y)$, we may identify $L^2(\mathbb{R}^2)$ with $L_x^2(\mathbb{R}; L_y^2(\mathbb{R}))$ and $\sigma(t)$ with the one-parameter shift of $L_x^2(\mathbb{R}; L_y^2(\mathbb{R}))$. As in Case A, if $\Gamma' \sim B'$ then $\Gamma' \colon L_x^2(\mathbb{R}; L_y^2(\mathbb{R})) \to L_x^2(\mathbb{R}; L_y^2(\mathbb{R}))$ commutes with the shift $\sigma(t)$ of $\mathbb{R}_x$, and $\Gamma'F = \pmb{\phi}_y F$, where this time, for each $x \in \mathbb{R}$, $F(x)$ is an element of $L_y^2(\mathbb{R})$ and $\pmb{\phi}_y(x)$ is an operator in $L_y^2(\mathbb{R})$. Thus

$$B'(f_1,f_2) = B'(F_1,F_2) = \int_{\mathbb{R}} <\Phi_y(x)F_1(x),\ F_2(x)>_{L^2_y} dx$$

$$\tag{4.2}$$

$$= \int_{\mathbb{R}} \int_{\mathbb{R}} \Phi_y(x)f_1(x,y)\overline{f_2(x,y)}\,dydx$$

and $\|\Phi_y(x)\| \le \|B\|$, $\forall x \in \mathbb{R}$. Since B' is also $\tau(t)$-invariant, it is easy to see that $\Phi_y: L^2_y \to L^2_y$ commutes with the shift in L^2_y, so that

$$\Phi_y(x)f(x,y) = \varphi'(x,y)f(x,y),\ \varphi' \in L^\infty(\mathbb{R}^2),\ \|\varphi'\|_\infty = \|B\|. \tag{4.3}$$

Thus, by (4.2) and (4.3),

$$B(f_1,f_2) = \iint f_1(x,y)\overline{f_2(x,y)}\,\varphi'(x,y)dxdy,\ \forall(f_1,f_2) \in W_1 \times W_2^\sigma. \tag{4.4}$$

Similarly,

$$B(f_1,f_2) = \iint f_1(x,y)\overline{f_2(x,y)}\ \varphi''(x,y)dxdy,\ \forall(f_1,f_2) \in W_1 \times W_2^\tau. \tag{4.4a}$$

A pair $\varphi',\varphi'' \in L^\infty(\mathbb{R}^2)$, satisfying (4.4) and (4.4a), will be called a *symbol of the first kind* of $B \sim \Gamma$. φ' and φ'' are functions, and they determine the form B in $W_1 \times W_2$, as well as the norm of $B\big|_{W_1 \times W_2^\sigma}$ and

the norm of $B\big|_{W_1 \times W_2^\tau}$, <u>but not</u> the norm of B in $W_1 \times W_2$, though by (3.2a),

they give an estimate of it.

On the other hand, if we consider B as only a $\tau(t)$-Hankel form (not necessarily $\sigma(t)$-Hankel) in the vector-valued space $L^2_x(\mathbb{R};\ L^2_y(\mathbb{R}))$, then Theorem 2.1 insures the existence of an operator-valued function, assigning to each $x \in \mathbb{R}$ an operator $\Psi'(x)$ in $L^2_y(\mathbb{R})$, such that

$$B(f_1,f_2) = \iint \Psi'(x)f_1(x,y)\overline{f_2(x,y)}\,dydx,\ \forall(f_1,f_2) \in W_1 \times W_2, \tag{4.5}$$

where

$$\|\Psi'\| = \sup\{\|\Psi'(x)\|:\ x \in \mathbb{R}\} = \|B\|. \tag{4.5a}$$

Similarly,

$$B(f_1,f_2) = \iint \Psi''(y)f_1(x,y)f_2(x,y)dxdy,\ \forall(f_1,f_2) \in W_1 \times W_2 \tag{4.5b}$$

where

$$\|\Psi''\| = \sup\{\|\Psi''(y)\|:\ y \in \mathbb{R}\} = \|B\|. \tag{4.5c}$$

This pair Ψ',Ψ'' will be called a *symbol of the second kind* of $B \sim \Gamma$. Each of the functions Ψ' and Ψ'' determine both the form <u>and</u> its norm in the whole of its domain $W_1 \times W_2$, but now, for each x or y, the operators $\Psi'(x)$, $\Psi''(y)$ are <u>not</u> given by multiplication by a function but are arbitrary operators. Thus, φ' and φ'' are scalar-valued functions, i.e., special operators in L^2_y, L^2_x, while Ψ' and Ψ'' are operator-valued functions, more difficult to deal with, but determining the norm of the whole B and not only of its restrictions to $W_1 \times W^\sigma_2$ and $W_1 \times W^\tau_2$.

   In the particular case when $W_1 = H^2(\mathbb{R}^2) = \{f \epsilon L^2(\mathbb{R}^2): \hat{f}(\xi,\eta) = 0$ if $\xi < 0$ or $\eta < 0\}$, $W_2 = W^\perp_1$,   (4.6)

i.e., where $\Gamma \sim B$ as in (2.1) is a so-called <u>big Hankel operator</u>, more can be said.

   First observe that in this case $W^\sigma_2 = \{f \epsilon L^2(\mathbb{R}^2): \hat{f}(\xi,\eta) = 0$ if $\eta > 0\}$ and $W^\tau_2 = \{f \epsilon L^2(\mathbb{R}^2): \hat{f}(\xi,\eta) = 0$ if $\xi > 0\}$.

   From the classical Riemann-Lebesgue Lemma in $L^2(\mathbb{R}^2)$, it follows that now W_2 satisfies the Riemann-Lebesgue condition (see Section 3), so that inequalities (3.9a) in Theorem 3.3 hold. In particular, <u>there are no compact big Hankel operator</u> acting in $L^2(\mathbb{R}^2)$.

         On the other hand, the Hankel form B can be considered as a $\tau(t)$-Hankel form acting in the operator-valued space $L^2_x(\mathbb{R}; L^2_y(\mathbb{R}))$, as well as a $\sigma(t)$-Hankel form in $L^2_y(\mathbb{R}; L^2_x(\mathbb{R}))$. If $B = B_n \sim \Gamma_n$ is a $\sigma(t)$-Hankel form, and rank $\Gamma_n \leq n$, then $L = \ker \Gamma_n \subset H^2(\mathbb{R}^2) \cong H^2_x(\mathbb{R}; H^2_y(\mathbb{R}))$ can be considered as a subspace of the vector-valued space $H^2_x(\mathbb{R}; H^2_y(\mathbb{R}))$, invariant under the $\sigma(t)$'s, and of codimension $\leq n$. Thus we have the situation as at the end of Section 2, only that now L, instead of being an invariant subspace of the scalar-valued $H^2_x(\mathbb{R})$, is such a subspace of the vector-valued $H^2_x(\mathbb{R}; H^2_y(\mathbb{R}))$. So that, again, $L = bH^2_x(\mathbb{R}; H^2_y(\mathbb{R}))$, but here $b = b_k$ instead of being a scalar Blaschke product is an n product of operator-valued Blaschke-Potapov factors (cf. [T], [CS4]), of the form $(x - \mu)(x - \bar{\mu})^{-1} |1 + \mu^2| (1 + \mu^2)^{-1} Q + (I - Q)$, Im $\mu > 0$ and Q an orthogonal projector in the space $H^2_y(\mathbb{R})$. Thus, $B = B_n \sim \Gamma_n$, rank $\Gamma_n \leq n$, has a symbol of second kind (Ψ'_n, Ψ''_n) such that $\Psi'_n = b^*_x \Psi'_a$ where here

$b_x = b(x)$ is an operator-valued Blaschke-Potapov n-product (cf. [T]) and Ψ_n', Ψ_a' are symbols of the second kind, with Ψ_a' such that

$$B_a'(f_1,f_2) = \iint \Psi_a'(x)f_1(x,y)\overline{f_2(x,y)}\,dydx = 0, \ \forall(f_1,f_2) \ \epsilon \ W_1 \times W_2. \quad (4.7)$$

Considering B as a $\tau(t)$-Hankel form acting on $L_y^2(\mathbb{R}; \ L_x^2(\mathbb{R}))$, every such $B = B_n \sim \Gamma_n$, rank $\Gamma_n \leq n$, has a symbol $\Psi_n'' = b_y^* \Psi_a''$, where again b_y as a function of y is a Blaschke-Potapov n-product and Ψ_a'' defines a zero form B_a'' in all $W_1 \times W_2$, as in (4.7). Thus equalities (3.6a) of Theorem 3.1 take the more precise form

$$s_n(\Gamma) = \inf\{\|\Psi_x - b_x^*\Psi_a'\|_\infty: \ b_x \text{ Blaschke-Potapov n-product}$$

$$\text{in the variable x, } \Psi_a'(x) \text{ defines zero } \tau\text{-Hankel form}\}$$

$$= \inf\{\|\Psi_y - b_y^*\Psi_a''\|_\infty: \ b_y \text{ Blaschke-Potapov n-product}$$

$$\text{in the variable y, } \Psi_a'' \text{ defines zero } \sigma\text{-Hankel form}\} \quad (4.8)$$

where Ψ_x and Ψ_y are operator-valued symbols of $B \sim \Gamma$ as a $\sigma(t)$-Hankel or a $\tau(t)$-Hankel form.

     This result, with obvious modifications, hold for any other pair of suitable subspaces W_1 and W_2, as in the case of $\mathbb{R}$ mentioned at the end of Section 2, replacing the Beurling-Lax Theorem by its vector-valued analogue (Halmos-Lax Theorem, cf. [N]).

5. THE A-A-K THEOREM IN THE REGULAR REPRESENTATION OF THE SYMPLECTIC PLANE.

     In this section we give the precise formulation of Theorems 3.2 and 3.3 in the case of the regular representation of $(\mathbb{R}^2, [\ ,\ ])$, using the conventions in [F].

     Let $(\mathbb{R}^2, [\ ,\ ]) \cong (\mathbb{C}, [\ ,\ ])$ be the symplectic plane, where the elements of $\mathbb{R}^2 \cong \mathbb{C}$ are written as $z = (x,y)$, and $[z_1,z_2] = \text{Im } z_1\bar{z}_2$ $= x_1y_2 - x_2y_1$. Consider the continuous 2-parameter evolution systems pair $[H_i; \ \sigma_i(t), \ \tau_i(t), \ t \ \epsilon \ \mathbb{R}]$, $i = 1,2$, where $H_1 = H_2 = L^2(\mathbb{R}^2)$, $\sigma_1 = \sigma_2 = \sigma$,

$\tau_1 = \tau_2 = \tau$ are the _twisted_ or _symplectic_ _shifts_ defined, for t,x,y ϵ ℝ, by

$$\sigma(t)f(x,y) = \exp(-\pi ity)f(x + t,y)$$

and   (5.1)

$$\tau(t)f(x,y) = \exp(\pi itx)f(x,t + y).$$

For every $\zeta = (s,t)$, $z = (x,y)$ set

$$W(\zeta)f(z) = \exp(-\pi i[z,\zeta])f(z + \zeta)$$ (5.2)

so that, for t ϵ ℝ,

$$\sigma(t) = W(t + i0), \quad \tau(t) = W(0 + it).$$ (5.2a)

Thus, σ and τ do not commute, but satisfy the Weyl commutation condition (3.1) with $a = \pi$, so the results of Section 3 apply here.

Now z $\longmapsto$ W(z) gives a unitary representation of the symplectic plane in $L^2(\mathbb{R}^2)$, since W(z) satisfy the Weyl-Segal relation

$$W(z)W(z') = \exp(\pi i[z,z']) W(z + z').$$ (5.2b)

This is called _the regular representation of_ $(\mathbb{C}, [,])$.

We consider here Hankel forms B: $W_1 \times W_2 \to \mathbb{C}$, where W_1, W_2 are subspaces of $L^2(\mathbb{R}^2)$ satisfying again (1.2), (1.2b) and (3.2a), and we concentrate below on the basic example where W_1 and W_2 are given by (4.6)

Let $L^2(L^2(\mathbb{R})$ be the space of all Hilbert-Schmidt operators T acting in $L^2(\mathbb{R})$, which is a Hilbert space under the scalar product

$$\langle T_1, T_2 \rangle = \operatorname{tr} T_2^* T_1 .$$ (5.3)

Every Hilbert-Schmidt operator T is given by a kernel k ϵ $L^2(\mathbb{R}^2)$,

$$T\varphi(x) = \int k(x,y)\varphi(y)dy, \quad \varphi \epsilon L^2(\mathbb{R}).$$ (5.4)

The **Weyl transform**, f $\mapsto$ W(f), establishes an isometric isomorphism between $L^2(\mathbb{R}^2)$ and $L^2(L^2(\mathbb{R}))$, by assigning to each f ϵ $L^2(\mathbb{R}^2)$ the Hilbert-Schmidt operator in $L^2(\mathbb{R})$ given by the kernel g ϵ $L^2(\mathbb{R}^2)$,

$$g(x,y) = g_f(x,y) = (F_2^{-1}f)(y - x, \frac{x + y}{2})$$ (5.5)

where F_2 is the Fourier transform in the second variable. The map f $\longmapsto$ g is clearly an isometric isomorphism of $L^2(\mathbb{R}^2)$ onto itself, so is f $\longmapsto$ W(f). Furthermore,

$$W(f) = \iint_{\mathbb{R}^2} f(x,y) \ w(-x,-y)dxdy$$ (5.6)

where

$$L^2(\mathbb{R}) \ni \varphi \mapsto (w(x,y)\varphi)(u) = \exp(2\pi iyu + \pi ixy)\varphi(u + x)$$ (5.7)

is the <u>Schrödinger unitary representation</u> of $(\mathbb{R}^2, [\ ,\ ])$ in $L^2(\mathbb{R})$.

Thus, under the isomorphism W of $L^2(\mathbb{R}^2)$ onto $L^2(L^2(\mathbb{R}))$, the operators $W(z)$ in (5.2) pass into the Schrödinger operators $w(z) = w(x,y)$, and $\sigma(t)$ and $\tau(t)$ into $w(t + i0)$ and $w(0 + it)$, respectively. Now $w(z)$ can be considered as an operator acting on $L^2(L^2(\mathbb{R}))$ by $A \longmapsto w(z)A$ for $A \in L^2(L^2(\mathbb{R}))$. Then, under the isomorphism W, the regular representation $\{W(z), z \in \mathbb{C}\}$ of the symplectic plane in $L^2(\mathbb{R}^2)$, passes into the unitary representation $\{w(z), z \in \mathbb{C}\}$ of the symplectic plane in $L^2(L^2(\mathbb{R}))$. Therefore, modulo a unitary isomorphism, the pair $[H_i;\ \sigma_i(t),\ \tau_i(t),\ t \in \mathbb{R}]$, $i = 1,2$, $H_1 = H_2 = L^2(\mathbb{R}^2)$, $\sigma_1 = \sigma_2 = \sigma$, $\tau_1 = \tau_2 = \tau$, the twisted shifts, is the same as the pair $[\tilde{H}_i; \tilde{\sigma}_i(t),\ \tilde{\tau}_i(t),\ t \in \mathbb{R}]$, $i = 1,2$, where $\tilde{H}_1 = \tilde{H}_2 = L^2(L^2(\mathbb{R}))$, $\tilde{\sigma}_1(t) = \tilde{\sigma}_2(t) = w(t + i0)$, $\tilde{\tau}_1(t) = \tilde{\tau}_2(t) = w(0 + it)$.

It is thus the same to study the Hankel or Toeplitz forms in the first or the second pair of systems. As shown in [CS1, Thm. V], every bounded Toeplitz form $B^\sim$ in the second pair of systems has the representation

$$B^\sim(T_1,T_2) = \text{tr } ST_2^*T_1 \tag{5.8}$$

where S is some bounded operator in $L^2(\mathbb{R})$. Therefore, by the lifting property in Section 3, for every bounded Hankel form B in the first pair of systems, there exist two bounded operators, S' and S'', acting in $L^2(\mathbb{R})$, such that

$$B(f_1,f_2) = \text{tr } S'W(f_2)^*\, W(f_1),\ \forall (f_1,f_2) \in W_1 \times W_2^\sigma$$

and $\hspace{8cm}$ (5.9)

$$B(f_1,f_2) = \text{tr } S''\, W(f_2)^*\, W(f_1),\ \forall (f_1,f_2) \in W_1 \times W_2^\tau.$$

The pair S',S'' will be called a *symbol of the first kind* of B. Here B is bounded in the $L^2(\mathbb{R})$-norm or, equivalently, $B^\sim$ is bounded in the Hilbert-Schmidt norm. If $B^\sim$ is in addition bounded in the operator norm, then the symbol operators S' and S'' are of trace class, given by <u>continuous</u> kernels $k'(x,y)$, $k''(x,y) \in L^2 \cap C$,

$$S' = W(k'),\quad S'' = W(k'') \tag{5.10}$$

and the kernels K' and K'', defined in $\mathbb{R}^2 \times \mathbb{R}^2$ by

$$K'(z,\zeta) = k'(z - \zeta)\ \exp(\pi i[z,\zeta])$$

and $\hspace{8cm}$ (5.10a)

$$K''(z,\zeta) = k''(z - \zeta)\ \exp(\pi i[z,\zeta])$$

give rise to positive definite matrix kernels $\begin{bmatrix} I & K' \\ K'^* & I \end{bmatrix}$, $\begin{bmatrix} I & K'' \\ K''^* & I \end{bmatrix}$.

Furthermore,

$$B(f_1,f_2) = \iint K'(z,\zeta)f_1(z)\overline{f_2(\zeta)} \, dz \, d\zeta \quad \text{in} \quad W_1 \times W_2^{\sigma}$$

and   (5.10b)

$$B(f_1,f_2) = \iint K''(z,\zeta)f_1(z)\overline{f_2(\zeta)} \, dz \, d\zeta \quad \text{in} \quad W_1 \times W_2^{\tau}.$$

On the other hand, if U_1 is the unitary isomorphism of $L^2(\mathbb{R}^2)$ onto itself defined by

$$(U_1f)(\zeta,y) = (F_1f)(\zeta + y/2,y),$$ (5.11)

the symplectic shift $\sigma(t)$ passes into the ordinary shift $f(x,y) \longmapsto e^{2\pi itx}f(x,y)$. Similarly, for

$$(U_2f)(x,\eta) = (F_2f)(x, \eta + \tfrac{x}{2}),$$ (5.11a)

the symplectic shift $\tau(t)$ passes into the ordinary shift $f(x,y) \longmapsto e^{2\pi ity}f(x,y)$.

Then, under U_1, every $\sigma(t)$-Hankel form $B: W_1 \times W_2 \to \mathbb{C}$ becomes a Hankel form $B_1: U_1W_1 \times U_1W_2 \to \mathbb{C}$ of the type considered in the case of $\mathbb{R}^2$, invariant under the ordinary shift in the operator-valued space $L_x^2(\mathbb{R}; L_y^2(\mathbb{R}))$, and thus, extends to a $\sigma(t)$-Toeplitz form in $L_x^2(\mathbb{R}; L_y^2(\mathbb{R}))$. Therefore B_1 is given by an operator-valued symbol $\phi_1 \in L_x^{\infty}(\mathbb{R}; L_y^2(\mathbb{R}))$. Similarly, every $\tau(t)$-Hankel $B: W_1 \times W_2 \longrightarrow \mathbb{C}$ becomes a Hankel $B_2: U_2W_1 \times U_2W_2 \longrightarrow \mathbb{C}$, invariant under $\tau(t)$, and extends to a $\tau(t)$-Toeplitz form, and has a symbol $\phi_2 \in L_y^{\infty}(\mathbb{R}; L_x^2(\mathbb{R}))$. Hence, since $B: W_1 \times W_2 \to \mathbb{C}$ is a Hankel form, it is both $\sigma(t)$- and $\tau(t)$-Hankel, so it is given by ϕ_1 as well as by ϕ_2, and the pair $(\phi_1,\phi_2) \in L_x^{\infty}(\mathbb{R}; L_y^2(\mathbb{R})) \times L_y^{\infty}(\mathbb{R}; L_x^2(\mathbb{R}))$ will be called a *symbol of the second kind* of B. By (4.5), (4.5b), for all $(f_1,f_2) \in W_1 \times W_2$,

$$B(f_1,f_2) = \iint \phi_1(x)(U_1f_1)(x,y)\overline{(U_1f_2)(x,y)}\,dxdy$$

(5.12)

$$= \iint \phi_2(y)(U_2f_1)(x,y)\overline{(U_2f_2)(x,y)}\,dydx$$

Moreover, since $U_1W_1 \subset L^2(\mathbb{R}^2) \sim L_x^2(\mathbb{R}; L_y^2(\mathbb{R}))$ is invariant under the ordinary shift $\sigma(t)$, it follows from the Halmos-Lax Theorem that

$$U_1 W_1 = \theta_{11} H^2_x(\mathbb{R}; E_1), \quad U_1 W_2 = \theta_{12} H^2_{-x}(\mathbb{R}; E_2)$$

where E_1, $E_2 \subset L^2_y(\mathbb{R})$, $\theta_{11}(x)$, $\theta_{12}(x) \in L(L^2_y(\mathbb{R}))$. Similarly,

$$U_2 W_1 = \theta_{21} H^2_y(\mathbb{R}; D_1), \quad U_2 W_2 = \theta_{22} H^2_{-y}(\mathbb{R}; D_2)$$

where $D_1, D_2 \subset L^2_x(\mathbb{R})$; $\theta_{21}(y)$, $\theta_{22}(y) \in L(L^2_x(\mathbb{R}))$. Hence, if B is Hankel of rank $\leq$ n, then B_1 is also of rank $\leq$ n, and therefore, as in the case of $\mathbb{R}^2$, it will be given by a symbol of the second kind, defined through Blaschke-Potapov operator-valued products.

It is easily seen that the Riemann-Lebesgue condition also holds in the symplectic case. Theorems 3.2 and 3.3 have in this case the following more precise version.

THEOREM 5.1. *Given a pair of continuous two-parameter evolution systems,* $[H_1; \sigma_1(t), \tau_1(t), t \in \mathbb{R}]$ *and* $[H_2; \sigma_2(t), \tau_2(t), t \in \mathbb{R}]$, *where* $H_1 = H_2 = L^2(\mathbb{R}^2)$, $\sigma_1 = \sigma_2 = \sigma$, $\tau_1 = \tau_2 = \tau$, *are the twisted shift (4.1), let* W_1 *and* $W_2 \subset L^2(\mathbb{R}^2)$ *be defined by (3.8), and let* $B: W_1 \times W_2 \to \mathbb{C}$ *be a bounded Hankel form,* $B^\sim \sim B$ *under the W isomorphism. Then*

(a) *B has a symbol of the first kind,* (S', S'') *such that the representations (5.9) hold in* $W_1 \times W^\sigma_2$, $W_1 \times W^\tau_2$. *Furthermore, if* $B^\sim$ *is bounded in the operator norm, then (5.10b) also hold.*

(b) *B has a symbol of the second kind,* (Ψ_1, Ψ_2) *such that the representation (5.12) hold in the whole of* $W_1 \times W_2$.

(c) *Relations (3.9a) of Theorem 3.3 hold, and, in particular, if B is compact,* $B \equiv 0$.

(d) *Relations (4.8) hold, where* (Ψ_x, Ψ_y) *is a symbol of second kind of B, and* $\Psi'_a \in L^\infty_x(\mathbb{R}; L^2_y(\mathbb{R}))$, $\Psi''_a \in L^\infty_y(\mathbb{R}; L^2_x(\mathbb{R}))$ *define zero* $\tau(t)$-*Hankel and* $\sigma(t)$-*Hankel forms, respectively.*

REMARK 5.1. As mentioned in the previous section, even if the symbols of the first kind don't give the norm or the singular numbers of the form, they still give an estimate of these numbers. Thus, an analogue to Theorem 5.1(d) can be given in terms of the symbols of the first kind, but within a fixed constant $\leq \sqrt{2}$.

REMARK 5.2. As mentioned above, every Hilbert-Schmidt operator $S \in L^2(L^2(\mathbb{R}))$ is the Weyl transform of a function $f \in L^2(\mathbb{R}^2)$. If $S \in L(L^2(\mathbb{R}))$ is only bounded, then S still gives rise to a Toeplitz form $B(f_1, f_2) = \text{tr } SW(f_2)^*W(f_1)$, and by Theorem 5.1(b), S is given by a pair of operator-valued functions $\Psi_1 \in L^2_y(\mathbb{R}; L^2_y(\mathbb{R}))$, $\Psi_2 \in L^2_y(\mathbb{R}; L^2_x(\mathbb{R}))$, which can be considered as a generalization of the Weyl transform to $L(L^2(\mathbb{R}))$.

REMARK 5.3. If $W(z)$, $z \in \mathbb{R}^2$, defined by (5.2), then the family C of all linear combinations $\Sigma_k c_k W(z_k)$ form a symmetric algebra of operators in $L^2(\mathbb{R}^2)$, and C generates a von Neumann algebra $A = C'' = (C')'$, where $C' = $ commutant of $C = \{A \in L(L^2(\mathbb{R}^2)): AT = TA, \forall T \in C\}$. If $B: L^2(\mathbb{R}^2) \times L^2(\mathbb{R}^2) \to C$ is a bounded form, and $B \sim \Gamma \in L(L^2(\mathbb{R}^2))$, then B is Toeplitz iff B is $W(z)$-invariant, $\forall z \in \mathbb{R}^2$, or, equivalently, if Γ commutes with $W(z)$, $\forall z \in \mathbb{R}^2$, i.e., if $\Gamma \in A'$. Thus (5.8) gives a description of the von Neumann algebra A'.

REMARK 5.4. If A is the algebra in the preceding remark, then, as a Banach space, A is the dual of a Banach space $A_* = $ the space of the linear forms T on A, $\langle T, X \rangle = \Sigma_{i=1}^\infty \langle \varphi_i, X\psi_i \rangle$, where $\Sigma\|\varphi_i\|^2 < \infty$, $\Sigma\|\psi_i\|^2 < \infty$. If $S \in A_*$ is of the form $\langle S, X \rangle = \Sigma_i \langle \varphi_i, X\varphi_i \rangle$, and $\|S\| = 1$, then S is called a <u>normal state</u> of A, and it is known that in the present case all the states S are <u>pure</u>, i.e., $\langle S, X \rangle = \langle \psi, X\psi \rangle$ for some $\psi \in L^2(\mathbb{R}^2)$, $\|\psi\| = 1$. Moreover, by a theorem of Irving Segal, a continuous function $F: \mathbb{R}^2 \to C$ satisfies $F(0) = 1$, $F(tz)$ is continuous in $t \in \mathbb{R}$ for all z, and

$$\Sigma_{jk} c_j \overline{c_k} \exp(\pi i [z_j, z_k]) F(z_j - z_k) \geq 0 \qquad (5.13)$$

iff $F(z) = W^{-1}(S)(z) = \text{tr } Sw(z)$ for some state S and all $z \in \mathbb{R}^2$. If P is the vector space of all the functions $\psi(z)$ of the type $\psi(z) = \Sigma_k c_k W(z_k)1(z)$, $1(z) \equiv 1$, then the sesquilinear forms $B: P \times P \to C$ which are $W(z)$-invariant and satisfy $B(\psi, \psi) \geq 0$, are in 1-1 correspondence with the functions F satisfying (5.13) and, therefore, also with the states S in A. Using this fact, the theory of Hankel forms in $L^2(\mathbb{R}^2, [,])$ developed above can be extended to Hankel forms in the space P, that are bounded with respect to a quadratic norm $\|\psi\|_S = B_S(\psi, \psi)^{1/2}$, where B_S is the positive form in P associated with the function $W^{-1}(S)(z)$, S a state in $\mathcal{A}$. Such Hankel forms in P can be used to give symplectic versions of classical harmonic analysis results.

REMARK 5.5. Theorem 5.1 is obtained by combining the Lifting theorem that extends Hankel to Toeplitz forms in general representations, with the Bochner integral representations of positive Toeplitz forms in regular representations. Integral representations can be given also in other unitary representations, provided they have a cyclic element or, more generally, a Lipshitz-Krein directing functional (cf. [CS2]) or for L^2-type representations (that can be injected in the regular representation). However, the commutative group $\mathbb{R}^n$ have no L^2-type representations, and its unitary representations need not be cyclic. Instead, L^2-type representations do exist for the Heisenberg group or the symplectic plane, and every representation of the symplectic plane has a cyclic element. This is one reason to study the Hankel forms in symplectic cases. Moreover, it is not hard to see that if there is a nontrivial symplectic representation in a Pontrjaguin space Π_κ, then $\kappa = 0$, i.e., Π_κ is Hilbert,

and, as observed in the remark above, the states of the associated von Neumann algebra will be of simple type or gaussian. Finally, in the symplectic case, such theories as the metaplectic representations, modular functions, Fock spaces of analytic functions or harmonic oscillators appear, which add interest to the study. A more detailed discussion of the symplectic case will be given in subsequent papers.

REFERENCES

[AAK]   V.M. Adamjan, V.Z. Arov and M.G. Krein, Analytic properties of Schmidt pairs of a Hankel operator and generalized Schur-Takagi problem, Mat. Sbornik 86(1971), 33-73.

[BH]   J. Ball and W. Helton, A Beurling-Lax theorem for the Lie group U(m,n) which contains most classical interpolation theory, J. Operator Theory 9(1983), 107-142.

[CS1]   M. Cotlar and C. Sadosky, Two-parameter lifting theorems and double Hilbert transforms in commutative and non-commutative settings, J. Math. Anal. & Appl. 150(1990), 439-480.

[CS2]   M. Cotlar and C. Sadosky, Toeplitz and Hankel forms related to unitary representations of the symplectic plane, Colloq. Math. **50/51** (1990), 639-708.

[CS3]   M. Cotlar and C. Sadosky, Transference of metrics induced by unitary couplings, J. Funct. Analysis <u>111</u> (1993), 473-488.

[CS4]   M. Cotlar and C. Sadosky, Abstract, Weighted and Multi-dimensional Adamjan-Arov-Krein theorems, and the singular numbers of Sarason commutants, Int. Eqs. & Op. Th. 17(1993).

[Da]   E.B. Davies, **One parameter semigroups**, Academic Press, New York, 1980.

[DS]   Dunford and J.T. Schwartz, **Linear Operators, I**, Interscience, New York, 1958.

[F]   Gerald B. Folland, **Harmonic analysis in phase space**, Princeton University Press, Princeton, 1989.

[H]   Henry Helson, **Lectures on invariant subspaces**, Academic Press, New York, 1964.

[N]   N.K. Nikolskii, **Treatise on the Shift Operator**, Springer-Verlag, Berlin, Heidelberg, New York, 1986.

[S]   D. Sarason, Generalized interpolation in H^∞, Trans. Amer. Math. Soc. **127**(1967), 179-203.

[Ta]   Michael Taylor, **Noncommutative Harmonic Analysis**, Amer. Math. Soc., Providence, 1986.

[T]   S. Treil, The theorem of Adamjan-Arow-Krein: Vector variant, Publ. Seminar LOMI Leningrad 141(1985), 56-72. (In Russian).

Mischa Cotlar
Department of Mathematics
Universidad Central de Venezuela
Caracas 1040, Venezuela

Cora Sadosky
Department of Mathematics
Howard University
Washington, DC 20059-USA

1991 Mathematics Subject Classification: 47B35, 43A65.

Operator Theory:
Advances and Applications, Vol. 71
© 1994 Birkhäuser Verlag Basel/Switzerland

PROJECTION METHOD FOR BLOCK TOEPLITZ OPERATORS WITH OPERATOR-VALUED SYMBOLS

I. Gohberg and M.A. Kaashoek

Dedicated to Harold Widom on the occasion of his 60-th birthday, with admiration and friendship

The convergence of the projection method for a block Toeplitz operator with a continuous operator-valued symbol is proved under the natural conditions on the operator involved. The result is based on a general abstract analysis of the convergence of projection methods which is also presented in this paper. For paired block Toeplitz operators a theorem about the convergence of the projection method is proved too.

0. INTRODUCTION

Let $T = [G_{i-j}]_{i,j=0}^{\infty}$ be a bounded block Toeplitz operator acting on $\ell_2(\mathcal{H})$, the Hilbert space of all square summable sequences with entries in the separable Hilbert space $\mathcal{H}$, and assume that the operator-valued symbol $G(z) = \sum_{j=-\infty}^{\infty} z^j G_j$ is continuous in the operator norm on the unit circle $\mathbb{T}$. One of the main results proved in this paper is the following theorem. If T and the associate block Toeplitz operator $\tilde{T} = [G_{j-i}]_{i,j=0}^{\infty}$ are invertible, then for N sufficiently large the finite section $T_N = [G_{i-j}]_{i,j=0}^{N-1}$ is an invertible operator on $\mathcal{H}^N$, the Hilbert space direct sum of N copies of $\mathcal{H}$, and for each $\mathbf{y} = (y_0, y_1, \ldots) \in \ell_2(\mathcal{H})$, we have

$$T^{-1}\mathbf{y} = \lim_{N \to \infty} (v_0, \ldots, v_{N-1}, 0, 0, \ldots),$$

with convergence in the norm of $\ell_2(\mathcal{H})$, where

$$\begin{pmatrix} v_0 \\ v_1 \\ \vdots \\ v_{N-1} \end{pmatrix} = T_N^{-1} \begin{pmatrix} y_0 \\ y_1 \\ \vdots \\ y_{N-1} \end{pmatrix}.$$

The invertibility conditions on T and $\tilde{T}$ are not only sufficient but also necessary. An analogous convergence result is proved for paired block Toeplitz operators $T = [T_{ij}]_{i,j=-\infty}^{\infty}$, where

$$T_{ij} = \begin{cases} G_{i-j}^{(1)}, & j \geq 0, \\ G_{i-j}^{(2)}, & j < 0, \end{cases}$$

such that $G^{(1)}(z) = \sum_{j=-\infty}^{\infty} z^j G_j^{(1)}$ and $G^{(2)}(z) = \sum_{j=-\infty}^{\infty} G_j^{(2)}(z)$ are continuous operator-valued functions on $\mathbb{T}$. These convergence results are obtained as corollaries of general abstract perturbation theorems. In the latter theorems the main novelty is that the perturbations are not required to be compact. We see these perturbation results as an operator modification and refinement of the original convergence proof of G. Baxter [Ba] (see also Section III.2 in [GF]).

Our investigation of the convergence of the projection method for the case of continuous operator-valued symbols was inspired by a recent preprint of A. Böttcher and B. Silbermann [BS2] in which this convergence was proved for another class of block Toeplitz operators with operator-valued symbols. In the Böttcher-Silbermann case the symbols are not necessarily continuous in the operator norm, but the perturbations involved turn out to be compact.

For Toeplitz and paired Toeplitz matrices with scalar or matrix entries our convergence results are well-known (see the books [GF] and [BS1] and the references therein). Our proofs have probably some new features for these cases too, and they can also be used in the non-Toeplitz case. In the usual proofs for the scalar and block matrix case compactness of the perturbations is used.

As one may expect (cf., [GK1] and [BS2]), the results of the present paper may be used to derive operator-valued versions of the Szegö-Widom limit formulas. We plan to return to these implications in a future publication.

This paper consists of five sections (not counting the present introduction). In the first section we derive conditions which guarantee that the projection method converges for a compression of an operator, assuming that this convergence is given for the full operator. The reverse implication is also considered. These results are used in the second section to prove our main convergence theorem for block Toeplitz operators with continuous operator-valued symbols. In Section 3 the results of the first section are extended to the case when the projections have limits into two directions. Section 4 gives the proof of the analogue of our main convergence theorem for paired block Toeplitz operators. In the last section we extend our convergence results for the projection method to an algebra of doubly infinite operator matrices (not of block Toeplitz type) which contains the block Laurent operators with continuous operator-valued symbols as a subalgebra.

Some words about notation. We denote by $\mathcal{L}(X, Y)$ the space of all bounded linear operators acting from the Banach space X into the Banach space Y. We write $\mathcal{L}(X)$ instead of $\mathcal{L}(X, X)$. The word subspace is used for a closed linear manifold in a Banach space. The algebraic direct sum of the two subspaces X_1, X_2 is denoted by $X_1 \dot{+} X_2$. The symbol $\ominus$ stands for an orthogonal direct sum in a Hilbert space.

We thank A. Böttcher and B. Silbermann for their comments on an earlier version of this paper.

1. PROJECTION METHOD FOR COMPRESSIONS

Throughout this section X and Y are complex Banach spaces, $\{P_\tau\}_{\tau \in \Omega}$ is a family of projections acting on X, and $\{Q_\tau\}_{\tau \in \Omega}$ is a family of projections acting on Y. Here Ω is a subset of $\mathbb{R}$ and $\sup \Omega = +\infty$. We assume that

$$(1.1) \qquad P_\tau x \to x \quad (x \in X), \qquad Q_\tau y \to y \quad (y \in Y)$$

for $\tau \to \infty$.

Let $A \in \mathcal{L}(X, Y)$ be given. We say that the *projection method relative to* $\{P_\tau, Q_\tau\}_{\tau \in \Omega}$ *is applicable to* A if beginning with some $\tau_0 \in \Omega$ for each $y \in Y$ the equation

$$(1.2) \qquad Q_\tau A P_\tau x = Q_\tau y \quad (\tau \geq \tau_0)$$

has a unique solution $x_\tau \in \operatorname{Im} P_\tau$ and as $\tau \to \infty$ the vectors x_τ tend to a solution of $Ax = y$. In this case we write $A \in \Pi\{P_\tau, Q_\tau\}_{\tau \in \Omega}$.

In the sequel A_τ denotes the operator from $\operatorname{Im} P_\tau$ into $\operatorname{Im} Q_\tau$ defined by

$$(1.3) \qquad A_\tau x = Q_\tau Ax \quad (x \in \operatorname{Im} P_\tau).$$

If the projection method is applicable to the operator A, then A is invertible, the operator A_τ is invertible for $\tau \geq \tau_0$ (for some $\tau_0 \in \Omega$) and, by the Banach-Steinhaus theorem,

$$(1.4) \qquad\qquad \sup_{\tau \geq \tau_0} \|A_\tau^{-1}\| < \infty.$$

The converse statement is also true, that is, if A is invertible and (1.4) holds (for some $\tau_0 \in \Omega$), then $A \in \Pi\{P_\tau, Q_\tau\}_{\tau \in \Omega}$. For these and related results the reader is referred to Section II.2 in [GF].

The operator $A \in \mathcal{L}(X, Y)$ is said to be a *compression* of the operator $\hat{A} \in \mathcal{L}(\hat{X}, \hat{Y})$, if X and Y are subspaces of the complex Banach spaces $\hat{X}$ and $\hat{Y}$, respectively, X has a closed linear complement X_1 in $\hat{X}$, the space Y has a closed linear complement Y_1 in $\hat{Y}$, and $\hat{A}$ partitions as

$$(1.5) \qquad\qquad \hat{A} = \begin{pmatrix} A & B \\ C & D \end{pmatrix} : X \dotplus X_1 \to Y \dotplus Y_1.$$

In the sequel we put

$$(1.6) \qquad\qquad \hat{P}_\tau = \begin{pmatrix} P_\tau & 0 \\ 0 & I_{X_1} \end{pmatrix}, \qquad \hat{Q}_\tau = \begin{pmatrix} Q_\tau & 0 \\ 0 & I_{Y_1} \end{pmatrix}.$$

Because of (1.1), we have

$$(1.7). \qquad\qquad \hat{P}_\tau \hat{x} \to \hat{x} \quad (\hat{x} \in \hat{X}), \qquad \hat{Q}_\tau \hat{y} \to \hat{y} \quad (\hat{y} \in \hat{Y})$$

for $\tau \to \infty$. In this section we prove the following two theorems.

THEOREM 1.1. *Let A be a compression of $\hat{A}$ as in (1.5), and let both A and $\hat{A}$ be invertible. Write*

$$(1.8) \qquad\qquad \hat{A}^{-1} = \begin{pmatrix} A^\times & B^\times \\ C^\times & D^\times \end{pmatrix} : Y \dotplus Y_1 \to X \dotplus X_1.$$

Assume that the projection method relative to $\{\hat{P}_\tau, \hat{Q}_\tau\}_{\tau \in \Omega}$ is applicable to $\hat{A}$. If, in addition,

$$(1.9) \qquad\qquad \lim_{\tau \to \infty} \|(I - P_\tau)B^\times\| = 0$$

or

(1.10) $$\lim_{\tau \to \infty} \|C^{\times}(I - Q_{\tau})\| = 0,$$

then $A \in \Pi\{P_{\tau}, Q_{\tau}\}_{\tau \in \Omega}$.

THEOREM 1.2. *Let A be a compression of $\hat{A}$ as in (1.5), and let both A and $\hat{A}$ be invertible. Partition $\hat{A}^{-1}$ as in (1.8), and assume that both (1.9) and (1.10) hold. Then for the projection method relative to $\{P_{\tau}, Q_{\tau}\}_{\tau \in \Omega}$ to be applicable to A it is necessary and sufficient that the projection method relative to $\{\hat{P}_{\tau}, \hat{Q}_{\tau}\}_{\tau \in \Omega}$ is applicable to $\hat{A}$.*

The two theorems above will be proved by repeatedly applying the following lemma.

LEMMA 1.3. *Let $A \in \mathcal{L}(X, Y)$ be invertible, and let $\{\Pi_{\tau}\}_{\tau \in \Lambda}$ and $\{\tilde{\Pi}_{\tau}\}_{\tau \in \Lambda}$ be uniformly bounded families of projections acting on X and Y, respectively. For each τ put:*

(1.11) $$A_{11}(\tau) = \tilde{\Pi}_{\tau} A \Pi_{\tau} | \text{Im } \Pi_{\tau} : \text{Im } \Pi_{\tau} \to \text{Im } \tilde{\Pi}_{\tau},$$

(1.12) $$A_{22}^{\times}(\tau) = (I - \Pi_{\tau}) A^{-1} (I - \tilde{\Pi}_{\tau}) | \text{Ker } \tilde{\Pi}_{\tau} : \text{Ker } \tilde{\Pi}_{\tau} \to \text{Ker } \Pi_{\tau}.$$

Then $A_{11}(\tau)$ is invertible for each $\tau \in \Lambda$ if and only if $A_{22}^{\times}(\tau)$ is invertible for each $\tau \in \Lambda$, and in this case there exists a constant γ (not depending on τ) such that

(1.13) $$\|A_{22}^{\times}(\tau)^{-1}\| \leq \gamma(1 + \|A_{11}(\tau)^{-1}\|), \quad \tau \in \Lambda,$$

(1.14) $$\|A_{11}(\tau)^{-1}\| \leq \gamma(1 + \|A_{22}^{\times}(\tau)^{-1}\|), \quad \tau \in \Lambda.$$

In particular, $A_{11}^{\times}(\tau)^{-1}$ is uniformly bounded with respect to τ if and only if $A_{22}^{\times}(\tau)^{-1}$ is uniformly bounded with respect to τ.

If $A_{11}(\tau)$ and $A_{22}^{\times}(\tau)$ are as in (1.11) and (1.12), respectively, then $A_{11}(\tau)$ and $A_{22}^{\times}(\tau)$ are said to be *matricially coupled* (the terminology is taken from [BGK]; see also [GGK], Section III.4), and in this case we refer to Π_{τ} and $\tilde{\Pi}_{\tau}$ as the *corresponding projections*.

PROOF OF LEMMA 1.3. Assume that $A_{11}(\tau)$ is invertible. Since $A_{11}(\tau)$ and $A_{22}^{\times}(\tau)$ are matricially coupled, we can apply Theorem I.2.1 in [BGK] (or Corollary

III.4.3 in [GGK]) to show that $A_{22}^{\times}(\tau)$ is invertible for each $\tau \in \Lambda$ and its inverse is the Schur complement of $A_{11}(\tau)$ in A, i.e.,

(1.15) $A_{22}^{\times}(\tau)^{-1} = A_{22}(\tau) - A_{21}(\tau)A_{11}(\tau)^{-1}A_{12}(\tau), \quad \tau \in \Lambda.$

Now, let $\tilde{m}$ be an upper bound for the norms $\|\tilde{\Pi}_\tau\|$ and $\|I - \tilde{\Pi}_\tau\|$ with $\tau \in \Lambda$ arbitrary. Then we see from (1.15) that

$$\|A_{22}^{\times}(\tau)^{-1}\| \leq \tilde{m}\|A\| + (\tilde{m}\|A\|)^2 \|A_{11}(\tau)^{-1}\|$$

for each $\tau \in \Lambda$. The latter inequality implies (1.13). The converse implication and (1.14) are proved in a similar way.   □

PROOF OF THEOREM 1.1. We shall assume that (1.9) holds; the proof for the case when (1.10) is satisfied follows the same line of reasoning as for (1.9). Consider the following partitioning of $\hat{A}$:

(1.16) $\hat{A} = \begin{pmatrix} E_\tau & F_\tau & G_\tau \\ H_\tau & A_\tau & B_\tau \\ K_\tau & C_\tau & D \end{pmatrix}$: Ker $P_\tau \dot{+}$Im $P_\tau \dot{+} X_1 \to$ Ker $Q_\tau \dot{+}$Im $Q_\tau \dot{+} Y_1.$

Here A_τ is precisely the operator defined by (1.3). We want to prove (1.4). Put

(1.17) $\hat{A}_\tau = \begin{pmatrix} A_\tau & B_\tau \\ C_\tau & D \end{pmatrix}$: Im $P_\tau \dot{+} X_1 \to$ Im $Q_\tau \dot{+} Y_1.$

Since $\hat{A} \in \Pi\{\hat{P}_\tau, \hat{Q}_\tau\}_{\tau \in \Omega}$ we know that there exists $\tau_0 \in \Omega$ such that $\hat{A}_\tau$ is invertible for $\tau \geq \tau_0$ and

(1.18) $\sup_{\tau \geq \tau_0} \|\hat{A}_\tau^{-1}\| < \infty.$

In what follows we take $\tau \geq \tau_0$.

We need to partition $\hat{A}^{-1}$ according to the partitioning of $\hat{A}$ in (1.16), as follows:

(1.19) $\hat{A}^{-1} = \begin{pmatrix} E_\tau^{\#} & F_\tau^{\#} & G_\tau^{\#} \\ H_\tau^{\#} & A_\tau^{\#} & B_\tau^{\#} \\ K_\tau^{\#} & C_\tau^{\#} & D^{\#} \end{pmatrix}$: Ker $Q_\tau \dot{+}$Im $Q_\tau \dot{+} Y_1 \to$ Ker $P_\tau \dot{+}$Im $P_\tau \dot{+} X_1.$

By comparing the partitionings of $\hat{A}^{-1}$ in (1.8) and (1.19) we see that $(I - P_\tau)B^\times = G_\tau^\#$ and hence (1.9) just means that

$$(1.20) \qquad \lim_{\tau \to \infty} \|G_\tau^\#\| = 0.$$

From (1.17) and the partitionings of $\hat{A}$ and $\hat{A}^{-1}$ in (1.16) and (1.19), respectively, we conclude that Lemma 1.3 applies to $E_\tau^\#$ and $\hat{A}_\tau$. Indeed, let Π_τ be the projection of $Y \dotplus Y_1$ onto $\operatorname{Ker} Q_\tau$ along $\operatorname{Im} Q_\tau \dotplus Y_1$, and let $\tilde{\Pi}_\tau$ be the projection of $X \dotplus X_1$ onto $\operatorname{Ker} P_\tau$ along $\operatorname{Im} P_\tau \dotplus X_1$. The convergencies in (1.1) imply that for $\tau \to \infty$ the operators Π_τ and $\tilde{\Pi}_\tau$ converge pointwise to the identity operators on $Y \dotplus Y_1$ and $X \dotplus X_1$, respectively. Hence, by the Banach-Steinhaus theorem, the families $\{\Pi_\tau\}_{\tau \in \Omega}$ and $\{\tilde{\Pi}_\tau\}_{\tau \in \Omega}$ are uniformly bounded. Now, apply Lemma 1.3 with $\hat{A}^{-1}$ in place of A. We conclude that $E_\tau^\#$ is invertible for $\tau \geq \tau_0$ and, because of (1.14), the inequality (1.18) yields

$$(1.21) \qquad \sup_{\tau \geq \tau_0} \|(E_\tau^\#)^{-1}\| < \infty.$$

Note that the 2×2 operator matrix in the left upper corner of $\hat{A}$ in (1.16) is just the operator A. Thus (1.16) and (1.19) show that A is matricially coupled to $D^\#$. According to our hypothesis A is invertible. Hence (cf., Lemma 1.3) we may conclude that $D^\#$ is invertible.

Next we show that

$$(1.22) \qquad \begin{pmatrix} E_\tau^\# & G_\tau^\# \\ K_\tau^\# & D^\# \end{pmatrix} : \operatorname{Ker} Q_\tau \dotplus Y_1 \to \operatorname{Ker} P_\tau \dotplus X_1$$

is invertible for τ sufficiently large. Indeed, note that (1.22) may be rewritten as a product $M_\tau N_\tau$, where

$$M_\tau = \begin{pmatrix} E_\tau^\# & 0 \\ K_\tau^\# & D^\# \end{pmatrix}, \qquad N_\tau = \begin{pmatrix} I_{\operatorname{Im} Q_\tau} & 0 \\ 0 & I_{Y_1} \end{pmatrix} + M_\tau^{-1} \begin{pmatrix} 0 & G_\tau^\# \\ 0 & 0 \end{pmatrix}.$$

From (1.21) and the invertibility of $D^\#$ it follows that M_τ is invertible and $\|M_\tau^{-1}\|$ is uniformly bounded on $\tau \geq \tau_0$. But then, using (1.20), we see that there exists $\tau_1 \geq \tau_0$ such that N_τ is invertible for $\tau \geq \tau_1$ and $\|N_\tau^{-1}\|$ is uniformly bounded on $\tau \geq \tau_1$. It follows that the operator in (1.22) is invertible for $\tau \geq \tau_1$ and

$$(1.23) \qquad \sup_{\tau \geq \tau_1} \left\| \begin{pmatrix} E_\tau^\# & G_\tau^\# \\ K_\tau^\# & D^\# \end{pmatrix} \right\| < \infty.$$

By interchanging rows and columns in (1.16) and (1.19) one sees that

$$(1.24) \qquad \begin{pmatrix} E_\tau & G_\tau & F_\tau \\ K_\tau & D & C_\tau \\ H_\tau & B_\tau & A_\tau \end{pmatrix}^{-1} = \begin{pmatrix} E_\tau^\# & G_\tau^\# & F_\tau^\# \\ K_\tau^\# & D^\# & C_\tau^\# \\ H_\tau^\# & B_\tau^\# & A_\tau^\# \end{pmatrix}.$$

Thus A_τ is matricially coupled to the operator defined by (1.22). Note that the corresponding projections are uniformly bounded because of (1.1). Thus we may use Lemma 1.3 and (1.23) to show that A_τ is invertible for $\tau \geq \tau_1$ and $\sup_{\tau \geq \tau_1} \|A_\tau^{-1}\| < 0$, which completes the proof. □

PROOF OF THEOREM 1.2. The sufficiency of the condition follows from Theorem 1.1. To prove the necessity we use the same line of arguments as in the proof of Theorem 1.1, but now in the reverse direction. Our aim is to show that (1.18) holds for some $\tau_0 \in \Omega$. Since $A \in \Pi\{P_\tau, Q_\tau\}_{\tau \in \Omega}$, we know that (1.4) is satisfied. From (1.16) and (1.19) we see that A_τ is matricially coupled to the operator defined by (1.22). Since the family of the corresponding projections are uniformly bounded, we may use Lemma 1.3 to show that (1.23) holds for some $\tau_1 \in \Omega$.

Next, note that (1.9) and (1.10) are equivalent to

$$(1.25) \qquad \lim_{\tau \to \infty} \|G_\tau^\#\| = 0, \qquad \lim_{\tau \to \infty} \|K_\tau^\#\| = 0,$$

respectively. We may write

$$\begin{pmatrix} E_\tau^\# & 0 \\ 0 & D^\# \end{pmatrix} = \begin{pmatrix} E_\tau^\# & G_\tau^\# \\ K_\tau^\# & D^\# \end{pmatrix} \left(I - \begin{pmatrix} E_\tau^\# & G_\tau^\# \\ K_\tau^\# & D^\# \end{pmatrix}^{-1} \begin{pmatrix} 0 & G_\tau^\# \\ K_\tau^\# & 0 \end{pmatrix} \right).$$

Thus (1.23) and (1.25) yield that $D^\#$ is invertible and there exists $\tau_2 \in \Omega$, $\tau_2 \geq \tau_1$, such that $E_\tau^\#$ is invertible for $\tau \geq \tau_2$ with

$$(1.26) \qquad \sup_{\tau \geq \tau_2} \|(E_\tau^\#)^{-1}\| < \infty.$$

Now, use that $E_\tau^\#$ is matricially coupled to $\hat{A}_\tau$, where $\hat{A}_\tau$ is defined by (1.17). The families of the corresponding projections are uniformly bounded. Thus we may use Lemma 1.3 and (1.26) to show that $\hat{A}_\tau$ is invertible for $\tau \geq \tau_2$ and (1.18) holds with $\tau_0 = \tau_2$. □

2. APPLICATIONS TO BLOCK TOEPLITZ OPERATORS WITH OPERATOR-VALUED SYMBOLS

Throughout this section $\mathcal{H}$ is a separable Hilbert space. By $\ell_2(\mathcal{H})$ we denote the Hilbert space of all square summable sequences $\mathbf{x} = (x_i)_{i=0}^{\infty}$ with elements in $\mathcal{H}$. A bounded linear operator T on $\ell_2(\mathcal{H})$ is called a $\mathcal{H}$-*block Toeplitz operator* if the action of T on $\ell_2(\mathcal{H})$ may be described as follows:

$$(2.1) \qquad (T\mathbf{x})_i = \sum_{j=0}^{\infty} G_{i-j} x_j, \qquad i = 0, 1, 2, \ldots,$$

where G_j ($j \in \mathbb{Z}$) is a bounded linear operator on $\mathcal{H}$. In this case the weakly measurable operator-valued function $G(\cdot)$ on the unit circle $\mathbb{T}$ whose j-th Fourier coefficient coincides with G_j ($j \in \mathbb{Z}$) is called the *defining function* of T. It is well-known (see [Be]) that

$$(2.2) \qquad \|T\| = \text{ess sup}\{\|G(\zeta)\| \mid \zeta \in \mathbb{T}\}.$$

Instead of (2.1) we shall write $T = [G_{i-j}]_{i,j=0}^{\infty}$.

Let $T = [G_{i-j}]_{i,j=0}^{\infty}$ be an $\mathcal{H}$-block Toeplitz operator. We set $\tilde{T} = [G_{j-i}]_{i,j=0}^{\infty}$, and call $\tilde{T}$ the *associate* of T. Note that $\tilde{T}$ is again an $\mathcal{H}$-block Toeplitz operator; its defining function $\tilde{G}(\cdot)$ is given by $\tilde{G}(z) = G(z^{-1})$, $z \in \mathbb{T}$.

Let n be an non-negative integer. By P_n we denote the orthogonal projection on $\ell_2(\mathcal{H})$ defined by

$$(2.3) \qquad P_n(x_0, x_1, x_2, \ldots) = (x_0, \ldots, x_n, 0, 0, \ldots).$$

Obviously, $P_n \mathbf{x} \to \mathbf{x}$ for each $\mathbf{x} \in \ell_2(\mathcal{H})$ whenever $n \to \infty$.

THEOREM 2.1. *Let $T = [G_{i-j}]_{i,j=0}^{\infty}$ be an $\mathcal{H}$-block Toeplitz operator with a continuous defining function $G(\cdot)$, and let P_n ($n \geq 0$) be as in (2.3). Then the projection method relative to $\{P_n, P_n\}_{n \geq 0}$ is applicable to T if and only if T and its associate $\tilde{T}$ are invertible operators on $\ell_2(\mathcal{H})$.*

For the proof of Theorem 2.1 we need some additional notation and a lemma. By $\ell_2^{\#}(\mathcal{H})$ we denote the Hilbert space consisting of all doubly infinite square summable sequences $\mathbf{x} = (x_j)_{j=-\infty}^{\infty}$ with elements in $\mathcal{H}$. With $\ell_2^{\#}(\mathcal{H})$ we associate the following two orthogonal projections:

$$(2.4) \qquad\qquad \mathbb{P}\big((x_j)_{j=-\infty}^{\infty}\big) = (\ldots, 0, 0, x_0, x_1, x_2, \ldots),$$

$$(2.5) \qquad\qquad \mathbb{Q}\big((x_j)_{j=-\infty}^{\infty}\big) = (\ldots, x_{-2}, x_{-1}, 0, 0, \ldots).$$

Obviously, $\mathbb{Q} = I - \mathbb{P}$. We say that L is an $\mathcal{H}$-block Laurent operator on $\ell_2^{\#}(\mathcal{H})$ if the action of L is given by

$$(2.6) \qquad\qquad (Lx)_i = \sum_{j=-\infty}^{\infty} G_{i-j} x_j, \qquad i \in \mathbb{Z},$$

where G_j ($j \in \mathbb{Z}$) is a bounded linear operator on $\mathcal{H}$. In this case $G(z) = \sum_{j=-\infty}^{\infty} z^j G_j$ is called the *defining function* of L.

LEMMA 2.2. *Let L be an $\mathcal{H}$-block Laurent operator on $\ell_2^{\#}(\mathcal{H})$ with a continuous defining function $G(\cdot)$. Then*

$$(2.7) \qquad\qquad \lim_{n \to \infty} \|(I - P_n^{\#})L\mathbb{Q}\| = 0,$$

$$(2.8) \qquad\qquad \lim_{n \to \infty} \|\mathbb{Q}L(I - P_n^{\#})\| = 0,$$

where $P_n^{\#}$ denotes the orthogonal projection on $\ell_2^{\#}(\mathcal{H})$ given by

$$(2.9) \qquad\qquad P_n^{\#}\big((x_j)_{j=-\infty}^{\infty}\big) = (\ldots, x_{n-2}, x_{n-1}, x_n, 0, 0, \ldots).$$

PROOF. We shall proof (2.7); the proof of (2.8) is similar. Take $\varepsilon > 0$. Since the defining function $G(\cdot)$ is continuous on $\mathbb{T}$, we can use the operator-valued version of the second Weierstrass approximation theorem to obtain an $\mathcal{L}(\mathcal{H})$-valued trigonometric polynomial $R(z) = \sum_{j=-p}^{q} z^j R_j$ such that

$$(2.10) \qquad\qquad \max_{z \in \mathbb{T}} \|G(z) - R(z)\| < \varepsilon.$$

(Note that the operator-valued version of the second Weierstrass approximation theorem may be proved by using the same arguments as in S.N. Bernstein's proof of the classical scalar version, which, for example, one may find in Sections 20 and 21 of [Ac]. We are indebted to Yu. Lubich for this remark.) Now let R be as in (2.10), and let L_R be the $\mathcal{H}$-block Laurent operator defined by R. Let $R_{ij}(n)$ denote the (i,j)-th entry in the doubly infinite operator matrix representation of $(I - P_n^\#)L_R\mathbb{Q}$. Then

$$R_{ij}(n) = \begin{cases} R_{i-j} & \text{if } i > n \text{ and } j < 0, \\ 0 & \text{otherwise,} \end{cases}$$

where R_{i-j} is the (i,j)-th entry in the operator matrix representation of L_R. But $R_{i-j} = 0$ for $i - j > p$. It follows that for each i, j the operator $R_{ij}(n) = 0$ whenever $n \geq p$, because in this case $i > n$ and $j < 0$ implies $i - j > p$. We conclude that $(I - P_N^\#)L_R\mathbb{Q} = 0$ for $n \geq p$.

Next, note that $\|(I - P_n^\#)L\mathbb{Q}\| \leq \|L - L_R\| + \|(I - P_n^\#)L_R\mathbb{Q}\|$, because of the orthogonality of the projections. By (2.10),

$$\|L - L_R\| = \max_{z \in \mathbb{T}} \|G(z) - R(z)\| < \varepsilon.$$

Hence $\|(I - P_n^\#)L\mathbb{Q}\| < \varepsilon$ for $n \geq p$, which proves the lemma.   □

PROOF OF THEOREM 2.1. We split the proof into three parts. In the first two parts we prove the sufficiency of the conditions.

Part (a). Let L be the $\mathcal{H}$-block Laurent operator defined by $G(\cdot)$, and let $P_n^\#$ be defined by (2.9). In this part we prove that L is invertible and $L \in \Pi\{P_n^\#, P_n^\#\}_{n \geq 0}$, whenever $\tilde{T}$ is invertible. Let $L_n : \operatorname{Im} P_n^\# \to \operatorname{Im} P_n^\#$ be defined by

$$L_n\mathbf{x} = P_n^\# L\mathbf{x}, \qquad \mathbf{x} \in \operatorname{Im} P_n^\#,$$

and let $U : \operatorname{Im} P_n^\# \to \ell_2(\mathcal{H})$ be the unitary operator given by

$$U(\ldots, x_{n-2}, x_{n-1}, x_n, 0, 0, \ldots) = (x_n, x_{n-1}, x_{n-2}, \ldots).$$

Then $UL_nU^{-1} = \tilde{T}$. Thus, if $\tilde{T}$ is invertible, then L_n is invertible and $\|L_n^{-1}\|$ does not depend on n. In particular,

(2.11) $$\sup_{n \geq 0} \|L_n^{-1}\| < \infty.$$

By Lemma III.1.1 in [GF] the latter implies that L is invertible. But, if L is invertible, then (see the fourth paragraph of Section 1 or Theorem II.2.1 in [GF]) formula (2.11) implies that $L \in \Pi\{P_n^\#, P_n^\#\}_{n \geq 0}$.

Part (b). In this part we assume that T and $\tilde{T}$ are invertible, and we show that $T \in \Pi\{P_n, P_n\}_{n\geq 0}$.

Let $\mathbb{P}$ and $\mathbb{Q}$ be the projections defined by (2.4) and (2.5). In the sequel we identify Im $\mathbb{P}$ with $\ell_2(\mathcal{H})$ by identifying the doubly infinite sequence $(\ldots 0, 0, x_0, x_1, x_2, \ldots)$ with the sequence $(x_0, x_1, x_2, \ldots)$. It follows that the $\mathcal{H}$-block Laurent operator L defined by $G(\cdot)$ admits the following partitioning

$$L = \begin{pmatrix} T & B \\ C & D \end{pmatrix} : \text{Im } \mathbb{P} \oplus \text{Im } \mathbb{Q} \to \text{Im } \mathbb{P} \oplus \text{Im } \mathbb{Q},$$

where T is the $\mathcal{H}$-block Toeplitz operator defined by $G(\cdot)$. Furthermore,

$$P_n^{\#} = \begin{pmatrix} P_n & 0 \\ 0 & I \end{pmatrix} : \text{Im } \mathbb{P} \oplus \text{Im } \mathbb{Q} \to \text{Im } \mathbb{P} \oplus \text{Im } \mathbb{Q}.$$

Thus we are in the position to apply the results of the previous section.

By the result of Part (a) we know that L is invertible and $L \in \Pi\{P_n^{\#}, P_n^{\#}\}_{n\geq 0}$. Let

$$L^{-1} = \begin{pmatrix} T^{\times} & B^{\times} \\ C^{\times} & D^{\times} \end{pmatrix}$$

be the partitioning of L^{-1} relative to the decomposition Im $\mathbb{P} \oplus$ Im $\mathbb{Q}$. By our hypothesis T is invertible. Thus (apply Theorem 1.1) in order to prove that $T \in \Pi\{P_n, P_n\}_{n\geq 0}$ it suffices to show that

$$(2.12) \qquad\qquad \lim_{n\to\infty} \|(I - P_n)B^{\times}\| = 0.$$

Note that the defining function of L^{-1} is given by $G(\cdot)^{-1}$, and this function is continuous. Hence, by Lemma 2.2, we have

$$\lim_{n\to\infty} \|(I - P_n^{\#})L^{-1}\mathbb{Q}\| = 0,$$

which, by the orthogonality of the projections, yields (2.12).

Part (c). In this part we assume that $T \in \Pi\{P_n, P_n\}_{n\geq 0}$. This implies that T is invertible. We have to prove that the same holds for $\tilde{T}$. To do this we use an argument of [GF] (see the proof of Theorem VIII.5.2 in [GF]). Let $E_n : \ell_2(H) \to \ell_2(H)$ be the operator defined by

$$E_n(x_0, x_1, x_2, \ldots) = (x_{n-1}, x_{n-2}, \ldots, x_0, x_n, x_{n+1}, \ldots).$$

Note that E_n is a unitary operator and $E_n = E_n^{-1}$. Now

$$P_n \tilde{T} P_n = E_n P_n T P_n E_n, \qquad n = 1, 2, \dots .$$

It follows that for n sufficiently large $\tilde{T}_n = P_n \tilde{T} | \mathrm{Im}\, P_n : \mathrm{Im}\, P_n \to \mathrm{Im}\, P_n$ is invertible and $\|\tilde{T}_n^{-1}\| = \|T_n^{-1}\|$. In particular, $\sup_{n \geq 0} \|\tilde{T}_n^{-1}\| < \infty$. But then we can apply Lemma III.1.1 in [GF] to show that $\tilde{T}$ is invertible. $\quad\square$

3. PROJECTION METHOD WITH TWO DIRECTIONS

Throughout this section A is a 2×2 operator matrix,

$$(3.1) \qquad A = \begin{pmatrix} A_{11} & A_{12} \\ A_{21} & A_{22} \end{pmatrix} : X_1 \oplus X_2 \to Y_1 \oplus Y_2,$$

whose entries are bounded linear operators acting between complex Banach spaces. Furthermore, for $\nu = 1, 2$, we have families $\{P_\nu(\tau)\}_{\tau \in \Omega}$ and $\{Q_\nu(\tau)\}_{\tau \in \Omega}$ of projections acting on X_ν and Y_ν, respectively. Here Ω is a subset of $\mathbb{R}$ (not depending on ν) such that $\sup \Omega = +\infty$, and we assume that

$$(3.2) \qquad P_\nu(\tau) x \to x \quad (x \in X_\nu), \qquad Q_\nu(\tau) y \to y \quad (y \in Y_\nu),$$

whenever $\tau \to \infty$.

In what follows we need the following projections

$$\hat{P}_1(\tau) = \begin{pmatrix} P_1(\tau) & 0 \\ 0 & I_{X_2} \end{pmatrix}, \quad \hat{P}_2(\tau) = \begin{pmatrix} I_{X_1} & 0 \\ 0 & P_2(\tau) \end{pmatrix}, \quad P(\tau) = \begin{pmatrix} P_1(\tau) & 0 \\ 0 & P_2(\tau) \end{pmatrix},$$

$$\hat{Q}_1(\tau) = \begin{pmatrix} Q_1(\tau) & 0 \\ 0 & I_{Y_2} \end{pmatrix}, \quad \hat{Q}_2(\tau) = \begin{pmatrix} I_{Y_2} & 0 \\ 0 & Q_2(\tau) \end{pmatrix}, \quad Q(\tau) = \begin{pmatrix} Q_1(\tau) & 0 \\ 0 & Q_2(\tau) \end{pmatrix}.$$

In this section we shall prove the following two theorems.

THEOREM 3.1. Let A in (3.1) be invertible, and assume that the projection methods relative to $\{\hat{P}_1(\tau), \hat{Q}_1(\tau)\}_{\tau \in \Omega}$ and $\{\hat{P}_2(\tau), \hat{Q}_2(\tau)\}_{\tau \in \Omega}$ are both applicable to A. Write

$$(3.3) \qquad A^{-1} = \begin{pmatrix} A_{11}^\times & A_{12}^\times \\ A_{21}^\times & A_{22}^\times \end{pmatrix} : Y_1 \dotplus Y_2 \to X_1 \dotplus X_2.$$

If, in additon,

(3.4)
$$\lim_{\tau \to \infty} \|(I - P_1(\tau))A_{12}^{\times}(I - Q_2(\tau))\| = 0$$

or

(3.5)
$$\lim_{\tau \to \infty} \|(I - P_2(\tau))A_{21}^{\times}(I - Q_1(\tau))\| = 0,$$

then $A \in \Pi\{P(\tau), Q(\tau)\}_{\tau \in \Omega}$.

THEOREM 3.2. *Let A in (3.1) be invertible, and partition A^{-1} as in (3.3). Assume that (3.4) and (3.5) are fulfilled. Then for the projection method relative to $\{P_\tau, Q(\tau)\}_{\tau \in \Omega}$ to be applicable to A it is necessary and sufficient that the projection methods relative to $\{\hat{P}_1(\tau), \hat{Q}_1(\tau)\}_{\tau \in \Omega}$ and $\{\hat{P}_2(\tau), \hat{Q}_2(\tau)\}_{\tau \in \Omega}$ are both applicable to A.*

PROOF OF THEOREM 3.1. We shall assume that (3.4) holds; the proof for the case when (3.5) is satisfied follows in a similar way. Let us partition A as follows:

(3.6)
$$A = \begin{pmatrix} E_{11}(\tau) & E_{12}(\tau) & E_{13}(\tau) & E_{14}(\tau) \\ E_{21}(\tau) & E_{22}(\tau) & E_{23}(\tau) & E_{24}(\tau) \\ E_{31}(\tau) & E_{32}(\tau) & E_{33}(\tau) & E_{34}(\tau) \\ E_{41}(\tau) & E_{42}(\tau) & E_{43}(\tau) & E_{44}(\tau) \end{pmatrix} :$$
$$\text{Ker } P_1(\tau) \dotplus \text{Im } P_1(\tau) \dotplus \text{Im } P_2(\tau) \dotplus \text{Ker } P_2(\tau) \to$$
$$\to \text{Ker } Q_1(\tau) \dotplus \text{Im } Q_1(\tau) \dotplus \text{Im } Q_2(\tau) \dotplus \text{Ker } Q_2(\tau).$$

Put

(3.7)
$$A(\tau) = \begin{pmatrix} E_{22}(\tau) & E_{23}(\tau) \\ E_{31}(\tau) & E_{33}(\tau) \end{pmatrix}.$$

We have to show that there exists $\tau_0 \in \Omega$ such that $A(\tau)$ is invertible for $\tau \geq \tau_0$ and

(3.8)
$$\sup_{\tau \geq \tau_0} \|A(\tau)^{-1}\| < \infty.$$

Since A is invertible, our assumption that the projection methods relative to $\{\hat{P}_1(\tau), \hat{Q}_1(\tau)\}_{\tau \in \Omega}$ and $\{\hat{P}_2(\tau), \hat{Q}_2(\tau)\}_{\tau \in \Omega}$ are applicable to A is equivalent to the statement that for some τ' and τ'' in Ω we have

(3.9)
$$\sup_{\tau \geq \tau'} \left\| \begin{pmatrix} E_{22}(\tau) & E_{23}(\tau) & E_{24}(\tau) \\ E_{32}(\tau) & E_{33}(\tau) & E_{34}(\tau) \\ E_{42}(\tau) & E_{43}(\tau) & E_{44}(\tau) \end{pmatrix}^{-1} \right\| < \infty$$

and

$$(3.10) \qquad \sup_{\tau \geq \tau''} \left\| \begin{pmatrix} E_{11}(\tau) & E_{12}(\tau) & E_{13}(\tau) \\ E_{21}(\tau) & E_{22}(\tau) & E_{23}(\tau) \\ E_{31}(\tau) & E_{32}(\tau) & E_{33}(\tau) \end{pmatrix}^{-1} \right\| < \infty.$$

To employ the latter two inequalities we need to partition A^{-1} according to the partioning of A in (3.6), as follows

$$(3.11) \qquad A^{-1} = \begin{pmatrix} E_{11}^{\times}(\tau) & E_{12}^{\times}(\tau) & E_{13}^{\times}(\tau) & E_{14}^{\times}(\tau) \\ E_{21}^{\times}(\tau) & E_{22}^{\times}(\tau) & E_{23}^{\times}(\tau) & E_{24}^{\times}(\tau) \\ E_{31}^{\times}(\tau) & E_{32}^{\times}(\tau) & E_{33}^{\times}(\tau) & E_{34}^{\times}(\tau) \\ E_{41}^{\times}(\tau) & E_{42}^{\times}(\tau) & E_{43}^{\times}(\tau) & E_{44}^{\times}(\tau) \end{pmatrix} :$$

$$\operatorname{Ker} Q_1(\tau) \dotplus \operatorname{Im} Q_1(\tau) \dotplus \operatorname{Im} Q_2(\tau) \dotplus \operatorname{Ker} Q_2(\tau) \rightarrow$$

$$\rightarrow \operatorname{Ker} P_1(\tau) \dotplus \operatorname{Im} P_1(\tau) \dotplus \operatorname{Im} P_2(\tau) \dotplus \operatorname{Ker} P_2(\tau).$$

From (3.6) and (3.11) it follows that $E_{11}^{\times}(\tau)$ is matricially coupled to $[E_{ij}(\tau)]_{i,j=2}^{4}$ and $E_{44}^{\times}(\tau)$ is matricially coupled to $[E_{ij}(\tau)]_{i,j=1}^{3}$. Note that the families of the corresponding projections are uniformly bounded, because of (3.2) and the Banach-Steinhaus theorem. Thus we may apply Lemma 3.1 to show that (3.9) and (3.10) imply that

$$(3.12) \qquad \sup_{\tau \geq \tau'} \|(E_{11}^{\times}(\tau))^{-1}\| < \infty,$$

$$(3.13) \qquad \sup_{\tau \geq \tau''} \|(E_{44}^{\times}(\tau))^{-1}\| < \infty.$$

By comparing the partitionings of A^{-1} in (3.3) and (3.11) one sees that $E_{14}^{\times}(\tau) = (I - P_1(\tau))A_{12}^{\times}(I - Q_2(\tau))$, and hence (3.4) is equivalent to the requirement that

$$(3.14) \qquad \lim_{\tau \to \infty} \|E_{14}^{\times}(\tau)\| = 0.$$

From the formulas (3.12), (3.13), and (3.14) we may conclude that for some $\tau_1 \in \Omega$ we have

$$(3.15) \qquad \sup_{\tau \geq \tau_1} \left\| \begin{pmatrix} E_{11}^{\times}(\tau) & E_{14}^{\times}(\tau) \\ E_{41}^{\times}(\tau) & E_{44}^{\times}(\tau) \end{pmatrix}^{-1} \right\| < 0.$$

In fact, to prove (3.15) one uses the same type of arguments as were used in Section 1 to prove formula (1.23).

Recall that $A(\tau)$ is given by (3.7). Hence we see from (3.6) and (3.11) that $A(\tau)$ is matricially coupled to the 2×2 operator matrix

(3.16)
$$\begin{pmatrix} E_{11}^{\times}(\tau) & E_{14}^{\times}(\tau) \\ E_{41}^{\times}(\tau) & E_{44}^{\times}(\tau) \end{pmatrix}.$$

Again (by (3.2)) the families of the corresponding projections are uniformly bounded. Hence we see from (3.15) by applying Lemma 1.3 that for $\tau \geq \tau_1$ the operator $A(\tau)$ is invertible and (3.8) holds with $\tau_0 = \tau_1$.   □

PROOF OF THEOREM 3.2. The sufficiency of the conditions follows from Theorem 3.1. To prove their necessity we use the same line of arguments as in the proof of Theorem 3.1, but now in the reverse direction.

Assume $A \in \Pi\{P(\tau), Q(\tau)\}_{\tau \in \Omega}$. Our aim is to show that (3.9) and (3.10) hold for some τ' and τ'' in Ω. According to our hypotheses (3.8) is satisfied for some $\tau_0 \in \Omega$. From (3.6), (3.7) and (3.11) we know that $A(\tau)$ is matricially coupled to the operator defined by (3.16). The families of the corresponding projections are uniformly bounded because of (3.2). Hence (3.8) and Lemma 1.3 imply that (3.15) holds for some $\tau_1 \in \Omega$.

Next, note that (3.4) and (3.5) are equivalent to

(3.17)
$$\lim_{\tau \to \infty} \| E_{14}^{\times}(\tau) \| = 0, \qquad \lim_{\tau \to \infty} \| E_{41}^{\times}(\tau) \| = 0,$$

respectively. We may write

$$\begin{pmatrix} E_{11}^{\times}(\tau) & 0 \\ 0 & E_{44}^{\times}(\tau) \end{pmatrix} =$$
$$= \begin{pmatrix} E_{11}^{\times}(\tau) & E_{14}^{\times}(\tau) \\ E_{41}^{\times}(\tau) & E_{44}^{\times}(\tau) \end{pmatrix} \left\{ I - \begin{pmatrix} E_{11}^{\times}(\tau) & E_{14}^{\times}(\tau) \\ E_{41}^{\times}(\tau) & E_{44}^{\times}(\tau) \end{pmatrix}^{-1} \begin{pmatrix} 0 & E_{14}^{\times}(\tau) \\ E_{41}^{\times}(\tau) & 0 \end{pmatrix} \right\}.$$

From this identity, (3.15) and (3.17) we see that there exists $\tau_2 \in \Omega$ such that

(3.18)
$$\sup_{\tau \geq \tau_2} \| E_{11}^{\times}(\tau)^{-1} \| < \infty, \qquad \sup_{\tau \geq \tau_2} \| E_{44}^{\times}(\tau)^{-1} \| < \infty.$$

Now, use that $E_{11}^{\times}(\tau)$ is matricially coupled to $[E_{ij}(\tau)]_{i,j=2}^4$ and $E_{44}^{\times}(\tau)$ is matricially coupled to $[E_{ij}(\tau)]_{i,j=1}^{\infty}$. The families of the corresponding projections are uniformly bounded. So (3.18) and Lemma 1.3 imply that (3.9) and (3.10) hold with $\tau' = \tau'' = \tau_2$.   □

One may prove somewhat stronger versions of the results obtained in this section by allowing the projections on X_2 and Y_2 to be parametrized by a set of indices different from the set of indices used to parametrize the projections on X_1 and Y_1. The fact that we used in this section a single index set Ω is not essential.

4. THE PROJECTION METHOD FOR BLOCK PAIR OPERATORS

Throughout this section $\mathcal{H}$ is a separable Hilbert space, and $\ell_2^\#(\mathcal{H})$ is the Hilbert space of all doubly infinlte square summable sequences with elements in $\mathcal{H}$. On $\ell_2^\#(\mathcal{H})$ we consider the operator

$$(4.1) \qquad\qquad A = R\mathbb{P} + S\mathbb{Q}.$$

Here R and S are $\mathcal{H}$-block Laurent operators defined by continuous $\mathcal{L}(\mathcal{H})$-valued functions on the unit circle, and $\mathbb{P}$ and $\mathbb{Q}$ are the orthogonal projections on $\ell_2^\#(\mathcal{H})$ given by (2.4) and (2.5), respectively. We shall refer to A in (4.1) as an $\mathcal{H}$-block pair operator (cf., [GF], Chapter V). In this section we are interested in the projection method for A relative to the orthogonal projections

$$(4.2) \qquad P_n\big((x_j)_{j=-\infty}^\infty\big) = (\ldots,0,0,x_{-n},\ldots,x_n,0,0,\ldots), \quad n = 1,2,\ldots.$$

We shall prove the following theorem.

THEOREM 4.1. *Let* $A = R\mathbb{P}+S\mathbb{Q}$ *be an $\mathcal{H}$-block pair operator, where R and S are invertible $\mathcal{H}$-block Laurent operators defined by continuous $\mathcal{L}(\mathcal{H})$-valued functions on the unit circle, and let P_n $(n \geq 1)$ be as in (4.2). Then the projection method relative to $\{P_n, P_n\}_{n\geq 1}$ is applicable to A if and if the operators*

$$(4.3) \qquad\qquad \mathbb{P} + \mathbb{Q}R\mathbb{Q}, \qquad \mathbb{P} + \mathbb{Q}R^{-1}S\mathbb{Q}, \qquad \mathbb{P}S\mathbb{P} + \mathbb{Q}$$

are invertible.

Let us consider the special case when $R = S$. Then A is just an $\mathcal{H}$-block Laurent operator defined by a continuous $\mathcal{L}(\mathcal{H})$-valued function, Φ say. In this case the three operators in (4.3) are invertible if and only if the $\mathcal{H}$-block Toeplitz operator T defined by Φ and its associate $\tilde{T}$ are invertible. It follows that Theorem 2.1 may be viewed as a particular case of Theorem 4.1.

To prove Theorem 4.1 we need the following intermediate result.

THEOREM 4.2. *Let* $A = R\mathbb{P} + S\mathbb{Q}$ *be an* $\mathcal{H}$-*block pair operator, where* R *and* S *are* $\mathcal{H}$-*block Laurent operators, and let* $P_n^{\#}$ *be the projection on* $\ell_2^{\#}(\mathcal{H})$ *defined by* (2.9). *Assume that* R *is invertible and defined by a continuous* $\mathcal{L}(\mathcal{H})$-*valued function on the unit circle. Then the projection method relative to* $\{P_n^{\#}, P_n^{\#}\}_{n \geq 1}$ *is applicable to* A *if and only if the operators*

$$(4.4) \qquad \mathbb{P} + \mathbb{Q}R\mathbb{Q}, \qquad \mathbb{P} + \mathbb{Q}R^{-1}S\mathbb{Q}$$

are invertible.

PROOF. From the invertibility of R we conclude that A may be rewritten as

$$(4.5) \qquad A = R(\mathbb{P} + R^{-1}S\mathbb{Q}).$$

It follows that A is invertible if and only if the second operator in (4.4) is invertible. Note that $A \in \Pi\{P_n^{\#}, P_n^{\#}\}_{n \geq 1}$, implies that A is invertible (cf., the fourth paragraph of Section 1). Therefore, in what follows we may assume without loss of generality that A is invertible.

Let $A_n = P_n^{\#} A | \mathrm{Im}\, P_n^{\#} : \mathrm{Im}\, P_n^{\#} \to \mathrm{Im}\, P_n^{\#}$. Since A is assumed to be invertible, we have $A \in \Pi\{P_n^{\#}, P_n^{\#}\}_{n \geq 1}$ if and only if there exists a positive integer n_0 such that A_n is invertible for $n \geq n_0$ and

$$(4.6) \qquad \sup_{n \geq n_0} \|A_n^{-1}\| < \infty.$$

Now write A, A^{-1}, R^{-1} and $\mathbb{P} + R^{-1}S\mathbb{Q}$ as 3×3 operator matrices relative to the orthogonal decomposition

$$\ell_2^{\#}(\mathcal{H}) = \mathrm{Im}\, \mathbb{Q} \ominus \mathrm{Im}\, (P_n^{\#} - \mathbb{Q}) \ominus \mathrm{Ker}\, P_n^{\#},$$

as follows,

$$A = \begin{pmatrix} A_{11} & A_{12}(n) & A_{13}(n) \\ A_{21}(n) & A_{22}(n) & A_{23}(n) \\ A_{31}(n) & A_{32}(n) & A_{33}(n) \end{pmatrix}, \quad A^{-1} = \begin{pmatrix} A_{11}^{\times} & A_{12}^{\times}(n) & A_{13}^{\times}(n) \\ A_{21}^{\times}(n) & A_{22}^{\times}(n) & A_{23}^{\times}(n) \\ A_{31}^{\times}(n) & A_{32}^{\times}(n) & A_{33}^{\times}(n) \end{pmatrix},$$

$$R^{-1} = \begin{pmatrix} R_{11}^{\times} & R_{12}^{\times}(n) & R_{13}^{\times}(n) \\ R_{21}^{\times}(n) & R_{22}^{\times}(n) & R_{23}^{\times}(n) \\ R_{31}^{\times}(n) & R_{32}^{\times}(n) & R_{33}^{\times}(n) \end{pmatrix},$$

$$(\mathbb{P} + R^{-1}S\mathbb{Q})^{-1} = \begin{pmatrix} G_{11} & 0 & 0 \\ G_{21}(n) & I_{\text{Im}\,(P_n^{\#}-Q)} & 0 \\ G_{31}(n) & 0 & I_{\text{Ker}\,P_n} \end{pmatrix}.$$

Note that

$$A_n = \begin{pmatrix} A_{11} & A_{12}(n) \\ A_{21}(n) & A_{22}(n) \end{pmatrix},$$

and this operator is matricially coupled to $A_{33}^{\times}(n)$. Since the corresponding projection $P_n^{\#}$ is orthogonal, we can apply Lemma 1.3 to show that (4.6) holds if and only if

(4.7)
$$\sup_{n \geq n_0} \|(A_{33}^{\times}(n))^{-1}\| < \infty.$$

We conclude that $A \in \Pi\{P_n^{\#}, P_n^{\#}\}_{n \geq 1}$ if and only if there exists a positive integer n_0 such that (4.7) is satisfied.

Next, we use (cf., formula (4.5)) that

(4.8)
$$A_{33}^{\times}(n) = G_{31}(n)R_{13}^{\times}(n) + R_{33}^{\times}(n).$$

Note that $\|G_{31}(n)\| \leq \|\mathbb{P} + R^{-1}S\mathbb{Q}\|$ for each n. Furthermore, $R_{13}^{\times}(n) = \mathbb{Q}R^{-1}(I - P_n^{\#})$. Since R is a $\mathcal{H}$-block Laurent operator defined by a continuous $\mathcal{L}(\mathcal{H})$-valued function on the unit circle, the same holds for R^{-1}. But then we can apply Lemma 2.2 to show that

$$\lim_{n \to \infty} \|R_{13}^{\times}(n)\| = 0.$$

We conclude that $G_{31}(n)R_{13}^{\times}(n) \to 0$ if $n \to \infty$.

Assume (4.7) holds. Then $(A_{33}^{\times}(n))^{-1}G_{31}(n)R_{13}^{\times}(n) \to 0$ if $n \to \infty$. It follows that for n sufficiently large the operator $R_{33}^{\times}(n)$ is invertible. But $R_{33}^{\times}(n)$ is matricially coupled to the operator

$$R(n) = P_n^{\#}R|\text{Im } P_n^{\#} : \text{Im } P_n^{\#} \to \text{Im } P_n^{\#}.$$

Thus $R(n)$ is invertible (cf., Lemma 1.3). Since R is a Laurent operator, $R(n)$ is unitarily equivalent to $R(-1)$. The latter operator is just $\mathbb{Q}R|\text{Im } \mathbb{Q}$. Thus $\mathbb{Q}R\mathbb{Q} + \mathbb{P}$ is invertible.

Conversely, assume $\mathbb{Q}R\mathbb{Q} + \mathbb{P}$ is invertible. Then $R(n)$ is invertible for each n and $\|R(n)^{-1}\|$ is independent of n. Recall that $R(n)$ and $R_{33}^{\times}(n)$ are matricially coupled. The corresponding projection is $P_n^{\#}$. Hence we can use Lemma 1.3 to show that $R_{33}^{\times}(n)$ is invertible for each $n \geq 0$ and

$$\sup_{n \geq 0} \|(R_{33}^{\times}(n))^{-1}\| < \infty.$$

This result, formula (4.8), and the fact that $G_{31}(n)R_{13}^{\times}(n) \to 0$ imply that (4.7) holds for some n_0.   □

PROOF OF THEOREM 4.1. By using arguments similar to the ones used in the first paragraph of the proof of Theorem 4.2 one sees that without loss of generality we may assume that the operators A and $\mathbb{P} + R^{-1}S\mathbb{Q}$ are invertible. In this case we have to show that $A \in \Pi\{P_n, P_n\}_{n \geq 1}$ if and only if $\mathbb{P} + \mathbb{Q}R\mathbb{Q}$ and $\mathbb{P}S\mathbb{P} + \mathbb{Q}$ are invertible operators. To do this we shall apply Theorem 3.2.

Write A^{-1} as a 2×2 operator matrix relative to the orthogonal decomposition Im $\mathbb{P} \oplus$ Im $\mathbb{Q}$, as follows

$$A^{-1} = \begin{pmatrix} A_{11}^{\times} & A_{12}^{\times} \\ A_{21}^{\times} & A_{22}^{\times} \end{pmatrix} : \text{Im } \mathbb{P} \oplus \text{Im } \mathbb{Q} \to \text{Im } \mathbb{P} \oplus \text{Im } \mathbb{Q}.$$

We need the following auxiliary projections:

$$P_1(n) = P_n|\text{Im } \mathbb{P} : \text{Im } \mathbb{P} \to \text{Im } \mathbb{P}, \quad P_2(n) = P_n|\text{Im } \mathbb{Q} : \text{Im } \mathbb{Q} \to \text{Im } \mathbb{Q}.$$

First we shall show that

$$(4.9) \qquad \lim_{n \to \infty} \|(I - P_2(n))A_{21}^{\times}(I - P_1(n))\| = 0,$$

$$(4.10) \qquad \lim_{n \to \infty} \|(I - P_1(n))A_{12}^{\times}(I - P_2(n))\| = 0.$$

Note (use (4.5)) that $A^{-1} = (\mathbb{P} + R^{-1}S\mathbb{Q})^{-1}R^{-1}$. Now

$$\mathbb{Q}(\mathbb{P} + R^{-1}S\mathbb{Q})^{-1} = \mathbb{Q}(\mathbb{P} + R^{-1}S\mathbb{Q})^{-1}\mathbb{Q},$$

and thus

$$\mathbb{Q}A^{-1}\mathbb{P} = \mathbb{Q}(\mathbb{P} + R^{-1}S\mathbb{Q})^{-1}\mathbb{Q}(\mathbb{Q}R^{-1}\mathbb{P}).$$

Since $A_{21}^\times$ is $\mathbb{Q}A^{-1}\mathbb{P}$ viewed as an operator from Im $\mathbb{P}$ into Im $\mathbb{Q}$, we have

(4.11)
$$\|(I - P_2(n))A_{21}^\times(I - P_1(n))\| \leq$$
$$\leq \|(I - P_2(n))\mathbb{Q}(\mathbb{P} + R^{-1}S\mathbb{Q})^{-1}\mathbb{Q}\|\,\|\mathbb{Q}R^{-1}\mathbb{P}(I - P_1(n))\|.$$

The first term in the right hand side of (4.11) is uniformly bounded in n. Since R is an $\mathcal{H}$-block Laurent operator defined by a continuous $\mathcal{L}(\mathcal{H})$-valued operator function on the unit circle, the same holds for R^{-1}. Thus we can apply Lemma 2.2 to show that the second term in the right hand side of (4.11) goes to zero if $n \to \infty$. This proves (4.9).

To prove (4.11) we write A as $A = S(S^{-1}R\mathbb{P} + \mathbb{Q})$. Since

$$\mathbb{P}(S^{-1}R\mathbb{P} + \mathbb{Q})^{-1} = \mathbb{P}(S^{-1}R\mathbb{P} + \mathbb{Q})^{-1}\mathbb{P},$$

it follows that $\mathbb{P}A^{-1}\mathbb{Q} = \mathbb{P}(S^{-1}R\mathbb{P} + \mathbb{Q})^{-1}\mathbb{P}(\mathbb{P}S^{-1}\mathbb{Q})$. Since S is an $\mathcal{H}$-block Laurent operator defined by a continuous $\mathcal{L}(\mathcal{H})$-valued operator function on the unit circle, the same holds for S^{-1}. Now apply Lemma 2.2 to S^{-1} with the projection $P_n^\#$ replaced by $P_n^\times$, where

(4.12) $$P_n^\times\left((x_j)_{j=-\infty}^\infty\right) = (\ldots, 0, 0, x_{-n}, x_{-n+1}, x_{-n+2}, \ldots).$$

It follows that $\lim_{n\to\infty} \|\mathbb{P}S^{-1}\mathbb{Q}(I - P_2(n))\| = 0$. By using the analogue of (4.11), with $A_{12}^\times$ in place of $A_{21}^\times$, we obtain in this way (4.10).

Consider the projections

$$\hat{P}_1(n) = P_n\mathbb{P} + \mathbb{Q}, \qquad \hat{P}_2(n) = \mathbb{P} + P_n\mathbb{Q}.$$

Since (4.9) and (4.10) are satisfied, we can apply Theorem 3.2 to show that $A \in \Pi\{P_n, P_n\}_{n\geq 1}$ if and only if the projection methods relative to $\{\hat{P}_1(n), \hat{P}_1(n)\}_{n\geq 1}$ and $\{\hat{P}_2(n), \hat{P}_2(n)\}_{n\geq 1}$ are applicable to A. Note that $\hat{P}_1(n)$ is the orthogonal projection $P_n^\#$ defined by (2.5) and $\hat{P}_2(n)$ is the orthogonal projection $P_n^\times$ given by (4.12). Since $\mathbb{P} + R^{-1}S\mathbb{Q}$ is assumed to be invertible, Theorem 4.2 shows that $A \in \Pi\{\hat{P}_1(n), \hat{P}_1(n)\}_{n\geq 1}$ if and only if $\mathbb{P} + \mathbb{Q}R\mathbb{Q}$ is invertible. By using the analogue of Theorem 4.2 for the projections $\hat{P}_2(n)$, $n \geq 1$, one proves that $A \in \Pi\{\hat{P}_2(n), \hat{P}_2(n)\}_{n\geq 1}$ if and only if $\mathbb{P}S\mathbb{P} + \mathbb{Q}$. Together these results complete the proof. □

5. THE THEOREM FOR THE NON-TOEPLITZ CASE

As before $\ell_2^{\#}(\mathcal{H})$ denotes the Hilbert space consisting of all doubly infinite square summable sequences with elements in $\mathcal{H}$. A bounded linear operator A on $\ell_2^{\#}(\mathcal{H})$ may be represented in a canonical way by a doubly infinite operator matrix $[A_{ij}]_{i,j=-\infty}^{\infty}$ whose entries A_{ij} are bounded linear operators on $\mathcal{H}$. In fact, we write $A = [A_{ij}]_{i,j=-\infty}^{\infty}$, whenever the action of A is given by

$$\left(A(x_j)_{j=-\infty}^{\infty} \right) = \left(\sum_{k=-\infty}^{\infty} A_{jk} x_k \right)_{j=-\infty}^{\infty}.$$

The operator A is called a *band operator* if there exists an integer N such that $A_{ij} = 0$ for $|i-j| > N$. In the sequel $\mathcal{A}$ denotes the smallest closed subalgebra of $\mathcal{L}(\ell_2^{\#}(\mathcal{H}))$ containing all band operators.

All $\mathcal{H}$-block Laurent operators defined by continuous $\mathcal{L}(\mathcal{H})$-valued operator functions are in $\mathcal{A}$. Indeed, if L is such an operator with defining function $G(\cdot)$, say, then given $\varepsilon > 0$ we may find (see the proof of Lemma 2.2) an $\mathcal{L}(\mathcal{H})$-valued trigonometric operator polynomial $F(\cdot)$ such that

$$\max_{z \in \mathbb{T}} \| G(z) - F(z) \| < \varepsilon.$$

Let L_F be the $\mathcal{H}$-block Laurent operator defined by $F(\cdot)$. Then L_F is a band operator and $\| L - L_F \| < \varepsilon$. Since $\varepsilon > 0$ is arbitrary, we see that $L \in \mathcal{A}$.

Also the $\mathcal{H}$-block pair operators $A = R\mathbb{P} + S\mathbb{Q}$ with R and S defined by continuous $\mathcal{L}(\mathcal{H})$-valued operator functions (see Section 4) are in $\mathcal{A}$. This follows from the result proved in the previous paragraph and the fact that $B\mathbb{P}$ is a band operator whenever B is. Finally, let us remark that also the nonstationary operator Wiener algebra (see [GK2], Section I.2, for the definition) is contained in $\mathcal{A}$.

In this section we consider the projection method for operators A from the algebra $\mathcal{A}$ relative to the orthogonal projections

(5.1) $P_n\left((x_j)_{j=-\infty}^{\infty} \right) = (\ldots, 0, 0, x_{-n}, \ldots, x_n, 0, 0, \ldots), \qquad n = 1, 2, \ldots.$

The main result is the following theorem

THEOREM 5.1. *Let A be from the algebra $\mathcal{A}$, and let P_n be the projection on $\ell_2^{\#}(\mathcal{H})$ given by (5.1). Assume A is invertible. Then the projection method relative to $\{P_n, P_n\}_{n \geq 1}$ is applicable to A if and only if $A \in \Pi\{\hat{P}_1(n), \hat{P}_1(n)\}_{n \geq 1}$ and $A \in \Pi\{\hat{P}_2(n), \hat{P}_2(n)\}_{n \geq 1}$, where*

(5.2) $$\hat{P}_1(n)(\dots, x_{-1}, x_0, x_1, \dots) = (\dots, x_{n-2}, x_{n-1}, x_n, 0, 0, \dots),$$

(5.3) $$\hat{P}_2(n)(\dots, x_{-1}, x_0, x_1, \dots) = (\dots, 0, 0, x_{-n}, x_{-n+1}, x_{-n+2}, \dots),$$

For the proof of the above theorem we need two lemmas.

LEMMA 5.2. *The algebra $\mathcal{A}$ is inverse closed, i.e., if $A \in \mathcal{A}$ and A is invertible as an operator on $\ell_2^{\#}(\mathcal{H})$, then $A^{-1} \in \mathcal{A}$.*

PROOF. Since $\mathcal{A}$ is a C*-subalgebra of the operator algebra $\mathcal{L}(\ell_2^{\#}(\mathcal{H}))$, the result follows from general Banach algebra theory. An alternative proof, which is of some independent interest goes as follows.

From the definition of $\mathcal{A}$ it follows that there exists a sequence $B_1, B_2, \dots$ of band operators such that $B_n \to A$ $(n \to \infty)$ in the norm of $\mathcal{L}(\ell_2^{\#}(\mathcal{H}))$. Since A is invertible (as an operator on $\ell_2^{\#}(\mathcal{H})$), we know that for n sufficiently large B_n is invertible and $B_n^{-1} \to A^{-1}$ $(n \to \infty)$ in the operator norm. But the algebra $\mathcal{A}$ is closed by definition. Hence it suffices to show that the inverse of a band operator is in $\mathcal{A}$.

So, let B be a band operator which is invertible (as an operator on $\ell_2^{\#}(\mathcal{H})$). Then the matrix entries of $B^{-1} = [B_{ij}^{\times}]_{i,j=-\infty}^{\infty}$ are exponentially decaying off the main diagonal, i.e., there exist constants $M \geq 0$ and $0 < a < 1$ such that

(5.4) $$\|B_{ij}^{\times}\| \leq M a^{|i-j|} \qquad (i, j \in \mathbb{Z}).$$

To see this we employ an argument which was used for the block matrix case in [BG]. Put $B(\lambda) = U(\lambda)BU(\lambda)^{-1}$, where for each $\lambda \in \mathbb{T}$ the operator $U(\lambda)$ is the diagonal operator on $\ell_2^{\#}(\mathcal{H})$ given by $U(\lambda)((x_j)_{j=-\infty}^{\infty}) = (\lambda^j x_j)_{j=-\infty}^{\infty}$. Since B is a band operator, $B(\lambda)$ is a trigonometric operator polynomial. Furthermore, because of the invertibility of B, the operator $B(\lambda)$ is invertible for each λ in some annulus $1 - \varepsilon < |\lambda| < 1 + \varepsilon$. Note that

$$B(\lambda)^{-1} = U(\lambda)B^{-1}U(\lambda)^{-1} = [\lambda^{i-j}B_{ij}^{\times}]_{i,j=-\infty}^{\infty}, \qquad \lambda \in \mathbb{T},$$

and the operator-valued function $B(\cdot)^{-1}$ is analytic on $1 - \varepsilon < |\lambda| < 1 + \varepsilon$. Moreover, the N-th coefficient in its Fourier expansion is the diagonal operator D_N given by

$$D_N\left((x_j)_{j=-\infty}^{\infty}\right) = (B_{j,j-N}^{\times} x_{j-N})_{j=-\infty}^{\infty}.$$

The usual estimates on the Fourier coefficients of an analytic function on $\mathbb{T}$ yield (5.4).

Now, let $C(N) = [C_{ij}(N)]_{i,j=-\infty}^{\infty}$ be the band operator defined by

$$C_{ij}(N) = \begin{cases} B_{ij}^{\times} & \text{if } |i - j| \leq N, \\ 0 & \text{otherwise.} \end{cases}$$

Then $C(N) \to B^{-1}$ $(N \to \infty)$ in the operator norm, because

$$\|C(N) - B^{-1}\| \leq \sum_{|k|>N} \sup_{|i-j|=k} \|B_{ij}^{\times}\| \leq M\left(\sum_{|k|>N} a^{|k|}\right) \to 0 \quad (N \to \infty).$$

Since, by definition, $C(N) \in \mathcal{A}$ for each N, we see that $B^{-1} \in \mathcal{A}$. $\square$

LEMMA 5.3. *Let $\mathbb{P}$ and $\mathbb{Q}$ be the orthogonal projections on $\ell_2^{\#}(\mathcal{H})$ given by (2.4) and (2.5), respectively. For each $A \in \mathcal{A}$ we have*

$$(5.5) \qquad \qquad \lim_{n \to \infty} \|(I - \hat{P}_1(n))A\mathbb{Q}\| = 0,$$

$$(5.6) \qquad \qquad \lim_{n \to \infty} \|(I - \hat{P}_2(n))A\mathbb{P}\| = 0.$$

Here $\hat{P}_1(n)$ and $\hat{P}_2(n)$ are defined by (5.2) and (5.3), respectively.

PROOF. We shall prove (5.5); the limit (5.6) may be established in a similar way. Let B a band operator. Since the projections $\mathbb{Q}$ and $\hat{P}_1(n)$ are orthogonal,

$$(5.7) \qquad \qquad \|(I - \hat{P}_1(n))A\mathbb{Q}\| \leq \|A - B\| + \|(I - \hat{P}_1(n))B\mathbb{Q}\|.$$

By definition, the band operators are dense in $\mathcal{A}$ with respect to the operator norm. Hence (5.7) shows that it suffices to prove the lemma for the case when A is a band operator.

Let $A = [A_{ij}]_{i,j=-\infty}^{\infty}$ be a band operator. Thus for some integer N we have $A_{ij} = 0$ whenever $|i - j| > N$. Let $T_{ij}(n)$ denote the (i,j)-th entry in the canonical operator matrix representation of $(I - \hat{P}_1(n))A\mathbb{Q}$. Then

$$(5.8) \qquad \qquad T_{ij}(n) = \begin{cases} A_{ij} & \text{if } i > n \text{ and } j < 0, \\ 0 & \text{otherwise.} \end{cases}$$

Note that $i > n$ and $j < 0$ imply that $|i - j| > n$. Since $A_{ij} = 0$ for $|i - j| > N$, it follows from (5.8) that $T_{ij}(n) = 0$ for each i and j whenever $n \geq N$. In other words, $(I - \hat{P}_1(n))A\mathbb{Q} = 0$ for $n \geq N$. In particular, (5.5) holds. $\square$

PROOF OF THEOREM 5.1. Introduce the following projections:

$$P_1(n) = P_n|\mathrm{Im}\ \mathbb{P} : \mathrm{Im}\ \mathbb{P} \to \mathrm{Im}\ \mathbb{P}, \quad P_2(n) = P_n|\mathrm{Im}\ \mathbb{Q} : \mathrm{Im}\ \mathbb{Q} \to \mathrm{Im}\ \mathbb{Q}, \quad n = 1, 2, \dots.$$

In the sequel we write operators on $\ell_2^\#(\mathcal{H})$ as 2×2 operator matrices relative to the decomposition $\mathrm{Im}\ \mathbb{P} \oplus \mathrm{Im}\ \mathbb{Q}$. Thus

$$\hat{P}_1(n) = \begin{pmatrix} P_1(n) & 0 \\ 0 & I_{\mathrm{Im}\ \mathbb{Q}} \end{pmatrix}, \quad \hat{P}_2(n) = \begin{pmatrix} I_{\mathrm{Im}\ \mathbb{P}} & 0 \\ 0 & P_2(n) \end{pmatrix}, \quad P(n) = \begin{pmatrix} P_1(n) & 0 \\ 0 & P_2(n) \end{pmatrix}.$$

According to our hypothesis A is invertible. We write

$$A^{-1} = \begin{pmatrix} A_{11}^\times & A_{12}^\times \\ A_{21}^\times & A_{22}^\times \end{pmatrix} : \mathrm{Im}\ \mathbb{P} \oplus \mathrm{Im}\ \mathbb{Q} \to \mathrm{Im}\ \mathbb{P} \oplus \mathrm{Im}\ \mathbb{Q}.$$

Note that

$$(I - \hat{P}_1(n))A^{-1}(I - \hat{P}_2(n)) = \begin{pmatrix} 0 & (I - P_1(n))A_{12}^\times(I - P_2(n)) \\ 0 & 0 \end{pmatrix}.$$

Furthermore, $\mathbb{Q}(I - \hat{P}_2(n)) = I - \hat{P}_2(n)$. Thus, by the orthogonality of the projections,

$$\|(I - P_1(n))A_{12}^\times(I - P_2(n))\| = \|(I - \hat{P}_1(n))A^{-1}(I - \hat{P}_2(n))\|$$
$$\leq \|(I - \hat{P}_1(n))A^{-1}\mathbb{Q}\|.$$

By Lemma 5.2 the operator A^{-1} is in the algebra $\mathcal{A}$. Thus we can apply Lemma 5.3 with A^{-1} in place of A. This yields (by (5.5)) that

$$\lim_{n \to \infty} \|(I - P_1(n))A_{12}^\times(I - P_2(n))\| = 0.$$

In a similar way (use (5.6) instead of (5.5)) one can show that

$$\lim_{n \to \infty} \|(I - P_2(n))A_{21}^\times(I - P_1(n))\| = 0.$$

In other words, formulas (3.4) and (3.5) are fulfilled in the present setting. So we can use Corollary 3.2 to finish the proof. $\square$

REFERENCES

[Ac] N.I. Achiezer, *Theory of approximation*, New York, 1956.

[Ba] G. Baxter, A norm inequality for a "finite section" Wiener-Hopf equation, *Illinois J. Math.* 7 (1963), 97-103.

[Be] H. Bercovici, *Operator theory and arithmetic in H_∞*, Math. Survey Monogr. 26, Amer. Math. Soc., Providence, R.I., 1988.

[BG] A. Ben-Artzi and I. Gohberg, Fredholm properties of band matrices and dichotomy, in: *Topics in operator theory, Constantin Apostol memorial issue*, (ed. I. Gohberg), Operator Theory: Advances and Applications 32, Birkhäuser Verlag, Basel, 1988, pp. 37-52.

[BGK] H. Bart, I. Gohberg and M.A. Kaashoek, The coupling method for solving integral equations, in: *Topics in operator theory, systems and networks, the Rehovot workshop*, (eds. H. Dym and I. Gohberg), Operator Theory: Advances and Applications 12, Birkhäuser Verlag, Basel, 1984, pp. 39-73. Addendum, *Integral Equations and Operator Theory* 8 (1985), 890-891.

[BS1] A. Böttcher and B. Silbermann, *Analysis of Toeplitz operators*, Springer Verlag, Berlin, 1990.

[BS2] A. Böttcher and B. Silbermann, Operator-valued Szegö-Widom limit theorems, this volume.

[GGK] I. Gohberg, S. Goldberg and M.A. Kaashoek, *Classes of Linear Operators*, Vol. I, Birkhäuser Verlag, Basel. 1990.

[GF] I.C. Gohberg and I.A. Fel'dman, *Convolution equations and projection methods for their solution*, Transl. Math. Monographs 41, Amer. Math. Soc., Providence R.I., 1974.

[GK1] I. Gohberg and M.A. Kaashoek, Asymptotic formulas of Szegö-Kac-Achiezer type, *Asymptotic Analysis* 5 (1992), 187-220.

[GK2] I. Gohberg and M.A. Kaashoek, The band extension on the real line as a limit of discrete band extensions, II. The entropy principle, in: *Continuous and discrete Fourier transforms, extension problems and Wiener-Hopf equations* (ed. I. Gohberg), Operator Theory: Advances and Applications 58, Birkhäuser Verlag, Basel, 1992; pp. 71-92.

I. Gohberg   M.A. Kaashoek
School of Mathematical Sciences   Faculteit Wiskunde en Informatica
Raymond and Beverley Sackler   Vrije Universiteit
Faculty of Exact Sciences   De Boelelaan 1081a
Tel Aviv University   1081 HV Amsterdam
69978 Tel-Aviv   The Netherlands
Israel

MSC 1991: 47B35

Operator Theory:
Advances and Applications, Vol. 71
© 1994 Birkhäuser Verlag Basel/Switzerland

SZEGÖ-WIDOM-TYPE LIMIT THEOREMS

Israel Gohberg and Naum Krupnik

*Dedicated to Harold Widom on his sixtieth birthday
with admiration and friendship*

This paper contains Szegö-Widom-type limit theorems for a new class of operators generated by Toepelitz and Hankel matrices. The traditional approach for Toeplitz operators is extended to this new class of operators.

1. Introduction.

This paper was motivated by some results from the following two publications [DG], [EG]. In the first one new limit formulas of Szegö-Widom-type were obtained. In the second a class of infinite matrices which behaves like finite Toeplitz matrices was analyzed. In the present paper we deduce limit theorems of Szegö-Widom-type for the class of matrices considered in [EG]. These theorems are generalizations of the theorems from [DG]. We also show that the traditional approach to Szegö-Widom-limit theorems presented in [W1,2] and [BS] can be extended to the case considered in this paper.

The paper consists of five sections and is organized as follows. In section 1 some known Szegö-Widom-limit theorems for Toeplitz operators and their perturbations by trace-class operators are presented. In section 2 we state several Szegö-Widom type limit theorems for a new class of matrices generated by Toeplitz and Hankel operators. A generalization of this class of matrices is studied in section 3. Some illustrative examples are considered in section 4. The proofs of the theorems stated in section 2 are given in section 5.

It is our pleasure to thank Bernd Silbermann for a useful discussion on the contents of this paper.

Preliminaries.

Let a be a continuous function on the unit circle $\mathbb{T}$ and

$$(1.1) \qquad a_k = \frac{1}{2\pi} \int_{\mathbb{T}} a(t) t^{-k} |dt|.$$

Denote by M the set of $a \in C(\mathbb{T})$ for which

$$(1.2) \qquad\qquad \sum_{m \in \mathbb{Z}} |ma_m^2| < \infty.$$

The set M was investigated by M.G. Krein [K]. The following two properties may be found in [K]:

1. The set M is a Banach algebra with the norm

$$(1.3) \qquad\qquad \|a\| := \left(\sum_{k \in \mathbb{Z}} |ma_m^2| \right)^{\frac{1}{2}} + \max_{t \in \mathbb{T}} |a(t)|.$$

2. The algebra M is inverse-closed in $C(\mathbb{T})$.

The latter means that any function $a \in M$ is invertible in M if and only if $a(t) \neq 0$ $(t \in \mathbb{T})$.

For any $a \in C(\mathbb{T})$ by $H(a)$ we denote the Hankel matrix $H(a) = [a_{j+k+1}]_{j,k=0}^{\infty}$. The condition (1.2) is equivalent to the condition where both the matrices $H(a)$ and $H(\tilde{a})$ $(\tilde{a}(t) = a(1/t))$ generate Hilbert-Schmidt operators in ℓ_2.

Denote by $\mathbb{K}$ the set of all operators A acting in ℓ_2 of the form $A = T(a) + K$, where $a \in M$, $T(a) = [a_{j-k}]_{j,k=0}^{\infty}$ is the Toeplitz operator and K is a trace class operator.

We will use the following properties of operators $A \in \mathbb{K}$.

3. Any operator $A \in \mathbb{K}$ admits a unique representation $A = T(a) + K$ where T is a compact operator and $a \in C(\mathbb{T})$. The proof can be found in [G].

4. The set $\mathbb{K}$ is a nonclosed subalgebra of the algebra $L(\ell_2)$ of linear bounded operators acting in ℓ_2.

This property follows from the well-known relation

$$(1.4) \qquad\qquad T(a)\, T(b) = T(ab) - H(a)\, H(\tilde{b}).$$

5. Algebra $\mathbb{K}$ is inverse-closed in $L(\ell_2)$.

Indeed, if $A = T(a) + K$ is invertible in $L(\ell_2)$, then $a(t) \neq 0$ $(t \in \mathbb{T})$ and

$$(1.5) \qquad\qquad A^{-1} = T(a^{-1}) + K_1,$$

where $K_1 = A^{-1}(I - T(a)\, T(a^{-1}) - KT(a^{-1}))$.

Denote by $M^{N \times N}$ the algebra of all matrices $[f_{jk}]_{j,k=1}^{N}$ with the entries f_{jk} from M. Let $a \in M^{N \times N}$ and $a_m = \int_{\mathbb{T}} a(t) t^{-m} |dt| (a_m \in \mathbb{C}^{N \times N})$.

Denote by $\mathbb{K}_N$ the algebra of all operators of the form $A = T(a) + K$, where $T(a) = [a_{j-k}]_{j,k=0}^{\infty}$ is the Toeplitz operator acting in the space

$$\ell_2(\mathbb{C}^N) = \left\{ \{\xi_k\}_{k=0}^{\infty} \mid \xi_k \in \mathbb{C}^n \;\; ; \;\; \sum_{k=0}^{\infty} \| \xi_k \|^2 < \infty \right\}$$

and K is a trace class operator in $\ell_2(\mathbb{C}^N)$.

It follows from properties 1 - 5 that $\mathbb{K}_N$ is an inverse-closed subalgebra of $L(\ell_2(\mathbb{C}^N))$ and for any $A = T(a) + K$ $(\in \mathbb{K}_N)$ its symbol $a = \operatorname{symb} A$ is uniquely defined by the operator A.

Let $A \in \mathbb{K}_N$, $a = \operatorname{symb} A$ $(in M^{N \times N})$, $\det a(t) \neq 0$ $(t \in \mathbb{T})$ and $\operatorname{ind} \det a = 0$.

Denote by $G(A)$ (and also by $G(a)$) the number

(1.6)
$$G(A) = \exp \frac{1}{2\pi} \int_{\mathbb{T}} \log \det a(t)|dt|.$$

For any operator

$$A = [a_{jk}]_{j,k=0}^{\infty} \quad (a_{jk} \in \mathbb{C}^{N \times N}, \; A \in \ell_2(\mathbb{C}^N))$$

we denote by A_n the matrix

(1.7)
$$A_n = [a_{jk}]_{j,k=0}^n.$$

Theorem 1.1. Let $A \in \mathbb{K}_N$ and $a = \operatorname{symb} A$. Suppose that $\det a(t) \neq 0$ $(t \in \mathbb{T})$ and the Toeplitz operator $T(a^{-1})$ is invertible in $\ell_2(\mathbb{C}^N)$. Then $T(a^{-1}) A - I$ is a trace class operator and

(1.8)
$$\lim_{n \to \infty} \frac{\det A_n}{G(A)^{n+1}} = \det(T(a^{-1})A)$$

where

(1.9)
$$G(A) = \exp \frac{1}{2\pi} \int_{\mathbb{T}} \log \det a(t)|dt| \quad (= G(a)).$$

For the case $A = T(a)$, this theorem was proved by H.Widom ([W1,2], see also [BS, ch. 10]). In the general case, it can be easily deduced from well-known results. Indeed, the operator A can be represented in the form $A = T^{-1}(a^{-1}) + K_1$ where K_1 is of trace class and, hence, it follows from [BS; 10.25, 10.26] that

$$\lim \frac{\det A_n}{G(a)^{n+1}} = \det (I + T(a^{-1})K_1) = \det (T(a^{-1})A).$$

Note that in the conditions of Theorem 1.1, the right part of (1.8) (as well as all the determinants $\det A_n$) may be zero.

However, if the operators $T(a^{-1})$ and A are invertible, then $\det (T(a^{-1})A) \neq 0$ and, hence, the following statement holds:

Theorem 1.2. *Let $A \in \mathbb{K}_N$ and $a = \operatorname{symb} A$. Suppose that $\det a(t) \neq 0$ $(t \in \mathbb{T})$ and the operators $T(a^{-1})$ and A are invertible in $\ell_2(\mathbb{C}^N)$. Then $\det A_n \neq 0$ for all sufficiently large n and*

$$(1.10) \qquad \lim_{n \to \infty} \frac{\det A_{n+1}}{\det A_n} = G(a).$$

§2. The main results.

In [EG], a class of infinite matrices

$$(2.1) \qquad B_n = \begin{bmatrix} u_0 & u_{-1} & \cdots & g_{n+1} & g_n \\ u_1 & u_0 & \cdots & g_{n+2} & g_{n+1} \\ \cdot & \cdot & \cdots & \cdot & \cdot \\ h_{-n-1} & h_{-n-2} & \cdots & v_0 & v_{-1} \\ h_{-n} & h_{-n-1} & \cdots & v_1 & v_0 \end{bmatrix}$$

that have many properties analogous to those of finite Toeplitz matrices is studied. In this section we state the limit Szegö-Widom-type theorems for such matrices.

Denote by B the following matrix.

$$(2.2) \qquad B = \begin{bmatrix} u_0 & u_{-1} & \cdots & g_1 & g_0 & g_{-1} & \cdots \\ u_1 & u_0 & \cdots & g_2 & g_1 & g_0 & \cdots \\ \cdot & \cdot & \cdots & \cdot & \cdot & \cdot & \cdots \\ h_{-1} & h_{-2} & \cdots & v_0 & v_{-1} & v_{-2} & \cdots \\ h_0 & h_{-1} & \cdots & v_1 & v_0 & v_{-1} & \cdots \\ h_1 & h_0 & \cdots & v_2 & v_1 & v_0 & \cdots \\ \cdot & \cdot & \cdots & \cdot & \cdot & \cdot & \cdots \end{bmatrix}$$

The matrix defined by (2.2) can be considered as an operator acting in the space $\mathcal{L} = \ell_2(\mathbb{C}^N) \times \tilde{\ell}_2(\mathbb{C}^N)$, where

$$\tilde{\ell}_2(\mathbb{C}^N) = \left\{ \{\xi_m\}_{m=-\infty}^{\infty} \mid \xi_m \in \mathbb{C}^N, \sum_{m=-\infty}^{\infty} \|\xi_m\|^2 < \infty \right\}.$$

Consider in the space $\mathcal{L}$ the sequence of projections

$$\hat{R}_n = \begin{bmatrix} I_N & 0 \\ 0 & \hat{P}_n \end{bmatrix} \qquad (n \in \mathbb{Z}),$$

where I_N is the identity operator in $\ell_2(\mathbb{C}^N)$ and

$$\hat{P}_n \{\xi_k\}_{k=-\infty}^{\infty} = (\ldots, \xi_{n-1}, \xi_n, 0, 0, \ldots).$$

It is clear that $B_n = \hat{R}_n B \hat{R}_n \mid \mathrm{Im}\ \hat{R}_n$. For any $n \in \mathbb{Z}$ we denote by $\tilde{B}_n$ the operator

$$(2.3) \qquad \tilde{B}_n = \hat{R}_n B \hat{R}_n + I - \hat{R}_n$$

Suppose that the operator B_1 is invertible in $\mathrm{Im}\ \hat{R}_1$, then the operator $\tilde{B}_1$ is invertible in $\mathcal{L}$ and $(\tilde{B}_1^{-1})_n B_n - \hat{R}_n$ is a finite-dimensional operator. We define

$$(2.4) \qquad d(B_n) := \det\left((\tilde{B}_1^{-1})_n\ B_n\right).$$

In order to state the first theorem, it is convenient to represent operator B, defined by (2.2) in the form

$$(2.5) \qquad B = \begin{bmatrix} B_1 & X \\ Y & T(v) \end{bmatrix},$$

where

$$X = \begin{bmatrix} g_0 & g_{-1} & \cdots \\ g_1 & g_0 & \cdots \\ \cdot & \cdot & \cdots \\ v_{-2} & v_{-3} & \cdots \\ v_{-1} & v_{-2} & \cdots \end{bmatrix}, \quad Y = \begin{bmatrix} h_0 & h_{-1} & \cdots & & \cdots & v_2 & v_1 \\ h_1 & h_0 & & \cdots & & v_3 & v_2 \\ \cdot & \cdot & & \cdots & \cdots & \cdot & \cdot \end{bmatrix}.$$

Theorem 2.1. *Let u, v, g, h belong to $M^{N \times N}$. Suppose that B_1 is invertible in $\mathrm{Im}\ \hat{R}_1$, $\det(v - hu^{-1}g)(t) \neq 0$ ($t \in \mathbb{T}$) and the matrix $w = v - hu^{-1}g$ admits a canonical factorization $w = w_+ w_-$. Then the operator $A = T(v) - YB_1^{-1}X$ belongs to the algebra $\mathbb{K}_N$, symb $A = w$ and*

$$(2.6) \qquad \lim_{n \to -\infty} \frac{d(B_n)}{G(w)^{1-n}} = \det(T(w^{-1})A).$$

If, moreover, the operator B is invertible in $\mathcal{L}$, then $d(B_n) \neq 0$ ($n < 0$) for sufficiently large $|n|$ and

$$(2.7) \qquad \lim_{n \to -\infty} \frac{d(B_n)}{d(B_{n+1})} = \exp\frac{1}{2\pi}\int_{\mathbb{T}} \log\det(v - hu^{-1}g)(t)|dt|.$$

Theorem 2.2. *Let $u, g, h, v \in M^{N \times N}$. Suppose that the operator B_1 is invertible in $\mathrm{Im}\ \hat{R}_1$ and the matrices u, v admit canonical factorizations $u = u_- u_+$, $v = v_+ v_-$. Then for sufficiently large n the determinants $d(B_n) \neq 0$,*

$$(2.8) \qquad \lim_{n \to +\infty} \frac{d(B_n)}{d(B_{n+1})} = G(B)$$

and

$$(2.9) \qquad \lim_{n \to +\infty} (d(B_n) \, G(B))^{n+1} = E(B),$$

where

$$(2.10) \qquad G(B) = \exp \frac{1}{2\pi} \int_{\mathbb{T}} \log \det v(t)|dt|$$

and

$$(2.11) \qquad E(B) = \det \left[T(\tilde{v}) \, (T(\tilde{v}) - H(\tilde{h}) \, T^{-1}(u) \, H(g))^{-1} \right].$$

The proofs of theorems 2.1, 2.2 are given in section 5.

We conclude this section with the following:

Theorem 2.3. *Let $g, h \in M^{N \times N}$ and $u = v = e$, where e is an identity matrix in $\mathbb{C}^{N \times N}$.
Suppose that the operator $I - H(h) \, H(g)$ is invertible in $\ell_2(\mathbb{C}^N)$, $\det(e - h(t)g(t)) \neq 0$ $(t \in \mathbb{T})$ and the matrix $w = e - hg$ admits a canonical factorization $w = w_+ \, w_-$. Then*
$$(2.12) \qquad \lim_{n \to -\infty} \frac{d(B_n)}{G(e - hg)^{1-n}} = \det \left[T \left((e - hg)^{-1} \right) \left(I - T(h)(I - H(\tilde{h}) \, H(g))^{-1} T(g) \right) \right].$$

If, moreover, the operator $I - T(h)(I - H(\tilde{h})H(g))^{-1}T(g)$ is invertible in $\ell_2(\mathbb{C}^N)$ then $d(B_n) \neq 0$ for all $n < 0$ with large enough $|n|$ and

$$(2.13) \qquad \lim_{n \to -\infty} \frac{d(B_{n-1})}{d(B_n)} = \exp \frac{1}{2\pi} \int_{\mathbb{T}} \log \det(e - h(t)g(t))|dt|.$$

This theorem for the case $h = g^*$ and $\|g\|_\infty < 1$ was established in [DG]. Note that in the paper [DG] the right part of (2.12) is represented in a different form:

$$(2.14) \qquad \lim_{n \to -\infty} \frac{d(B_n)}{G(e - g^*g)^{1-n}} = \frac{\det(I - H(g^*)H(g))}{\prod_{m=1}^{\infty} \left\{ \det(E - K_{-m}K_{-m}^*)^m \right\}}$$

where E is an identity matrix in $\mathbb{C}^{N \times N}$. In order to explain the meaning of the matrices K_{-m} we recall the following statement, which is proved in the paper [AAK].

Let the Hankel matrix

$$\Gamma_{j+1}(g) = \begin{bmatrix} g_{j+1} & g_{j+2} & \cdot & \cdot & \cdot \\ g_{j+2} & g_{j+3} & \cdot & \cdot & \cdot \\ \cdot & & \cdot & \cdot & \cdot \end{bmatrix}$$

be contractive, then Γ_j is contractive iff the new entry g_j is of the form

(2.15)
$$g_j = Z_{j+1} + L_{j+1}^{1/2} \, K_j \, M_{j+1}^{1/2},$$

where Z_{j+1}, L_{j+1}, and M_{j+1}, are defined by the entries of Γ_{j+1} by the following relations:

$$L_{j+1} = \Pi^* \left(I - \Gamma_{j+1} \, \Gamma_{j+1}^*\right)^{-1} \Pi,$$
$$M_{j+1} = \Pi^* \left(I - \Gamma_{j+1}^* \, \Gamma_{j+1}\right)^{-1} \Pi$$

and

$$Z_{j+1} = -\Pi^* \, \Gamma_{j+1} \, \Gamma_{j+2}^* \left(I - \Gamma_{j+1} \, \Gamma_{j+1}^*\right)^{-1} \Gamma_{j+1} \, \Pi \, M_{j+1},$$

where Π^* denotes the projection of $\xi \in \ell_2(\mathbb{C}^N)$ onto its first component $\xi_0 \in \mathbb{C}^N$. The matrices K_j are contractions uniquely defined by Γ_j in (2.15).

§3. Limit theorems for the algebra $\mathbb{K}_N^{m \times m}$.

Let A_{jk} $j, k = 1, \ldots, m$ be some linear bounded operators acting in $\ell_2(\mathbb{C}^N)$ and

$$P_n \{\xi_k\}_{k=0}^\infty = (\xi_0, \xi_1, \ldots, \xi_n, 0, 0, \ldots) \quad (\xi_k \in \mathbb{C}^N).$$

Consider in the space $\ell_2^m(\mathbb{C}^N)$ $(m > 1)$ the operator $A = [A_{jk}]_{j,k=1}^m$ and the sequence of projections

(3.1)
$$R_n = \begin{bmatrix} I_N & 0 & \cdots & 0 & 0 \\ 0 & I_N & \cdots & 0 & 0 \\ \cdot & \cdot & \cdots & & \cdot \\ 0 & 0 & \cdots & I_N & 0 \\ 0 & 0 & \cdots & 0 & P_n \end{bmatrix},$$

where I_N is the identity operator in the space $\ell_2(\mathbb{C}^N)$. We represent the operator A in the form

(3.2)
$$A = \begin{bmatrix} F & C \\ B & A_{mm} \end{bmatrix},$$

where

$$F = [A_{jk}]_{j,k=1}^{m-1},$$

(3.3)
$$B = [A_{m1}, \ldots, A_{m,m-1}], \quad C = [A_{1m}, \ldots, A_{m-1,m}]^T.$$

Denote by A_n and $\tilde{A}_n$ the following operators

$$A_n = R_n \, A \, R_n \,|\, \mathrm{Im}\, R_n$$

and

$$\tilde{A}_n = R_n A R_n + (I - R_n).$$

In particular,

$$F = [A_{jk}]_{j,k=1}^{m-1}, \quad \tilde{F} = \begin{bmatrix} F & 0 \\ 0 & I_N \end{bmatrix}.$$

We suppose that the operator F is invertible in $\ell_2^{m-1}(\mathbb{C}^N)$. In this case, the operator $(\tilde{F})_n^{-1}$ is of the form

$$\tilde{F}_n^{-1} = \begin{bmatrix} F^{-1} & 0 \\ 0 & E \end{bmatrix}$$

where E is an identity matrix.

It is evident that the operator $\tilde{F}_n^{-1} A_n$ which acts in $\operatorname{Im} R_n$ has the form $I + K$, where K is a finite-dimensional operator. We denote by

(3.4) $$\tilde{D}_n(A) = \det\left(\tilde{F}_n^{-1} A_n\right).$$

Lemma 3.1. Let $A \in \ell_2^m(\mathbb{C}^n)$ and let $\{R_n\}$ be the sequence of projections, defined by (3.1). Then

(3.5) $$\det(\tilde{F}_n^{-1} A_n) = \det(\tilde{F}^{-1} A)_n$$

and, hence, $\tilde{D}_n(A)$ can be defined by the equality

(3.6) $$\tilde{D}_n(A) = \det(\tilde{F}^{-1} A)_n.$$

Proof. It follows from the representation (3.2) and definition (3.4) that

(3.7) $$R_n \tilde{F}^{-1} R_n = \begin{bmatrix} F^{-1} & 0 \\ 0 & P_n \end{bmatrix}, \quad R_n A R_n = \begin{bmatrix} F & C P_n \\ P_n B & P_n A_{mm} P_n \end{bmatrix},$$

$$R_n \tilde{F}^{-1} A R_n = \begin{bmatrix} I & F^{-1} C P_n \\ P_n B & P_n A_{mm} P_n \end{bmatrix}$$

and, hence, $R_n \tilde{F}^{-1} R_n A R_n = R_n \tilde{F}^{-1} A R_n$.

Theorem 3.2. Let $A = [A_{jk}]_{j,k=1}^m \in L(\ell_2^m(\mathbb{C}^N))$ and let the operator $F = [A_{jk}]_{j,k=1}^{m-1}$ be invertible in $\ell_2^{m-1}(\mathbb{C}^N)$. Then

(3.8) $$\tilde{D}_n(A) = \det S_n$$

where
$$S = A_{mm} - B \, F^{-1} \, C \quad (\in L(\ell(\mathbb{C}^N)))$$

and the operators B and C are defined by (3.3).

Proof. It follows from (3.6), (3.7) that

(3.9) $\qquad \tilde{D}_n(A) = \det \begin{bmatrix} I & F^{-1} \, C \, P_n \\ P_n \, B & P_n \, A_{mm} \, P_n + Q_n \end{bmatrix} \quad (Q_n = I_N - P_n) \, .$

Now we use the Schur decomposition

(3.10) $\qquad \begin{bmatrix} I & F^{-1} \, C \, P_n \\ P_n \, B & P_n \, A_{mm} \, P_n + Q_n \end{bmatrix} =$

$$= \begin{bmatrix} I & 0 \\ P_n \, B & I \end{bmatrix} \begin{bmatrix} I & 0 \\ 0 & P_n(A_{mm} - B \, F^{-1} \, C)P_n + Q_n \end{bmatrix} \begin{bmatrix} I & F^{-1} \, C \, P_n \\ 0 & I \end{bmatrix} .$$

Since P_n are finite-dimensional projections it follows from (3.9), (3.10) that

$$\tilde{D}_n(A) = \det \, (A_{mm} - B \, F^{-1} \, C)_n = \det S_n,$$

and the theorem is proved.

Theorem 3.3. *Let the operator* $A \in \mathbb{K}_N^{m \times m}$ *be represented in the form* (2.2):

$$A = \begin{bmatrix} F & C \\ B & A_{mm} \end{bmatrix},$$

and let the matrix functions f, c, b, a_{mm} *and* s *be the symbols of the operators* F, C, B, A_{mm}
and

$$S = A_{mm} - B \, F^{-1} \, C$$

respectively.

Suppose that the operator F *is invertible in* $\ell_2^{m-1}(\mathbb{C}^N)$, *the matrix-function* $s(t)$ *is
nonsingular on* $\mathbb{T}$ ($\det s(t) \neq 0, \; t \in \mathbb{T}$) *and the operator* $T(s^{-1})$ *is invertible in* $\ell_2(\mathbb{C}^N)$.
    Then

(3.11) $\qquad\qquad\qquad \lim_{n \to \infty} \dfrac{\tilde{D}_n(A)}{G(A)^{n+1}} = \det(T(s^{-1})S)$

where the determinant $\tilde{D}_n(A)$ *is defined by* (3.4) *and*

(3.12) $\qquad\qquad G(A) = \exp \dfrac{1}{2\pi} \int_{\mathbb{T}} \log \dfrac{\det(\operatorname{symb} A)(t)}{\det(\operatorname{symb} F)(t)} \, |dt|.$

Proof. In theorem 3.2 we showed that $\tilde{D}_n(A) = \det(A_{mm} - BF^{-1} C)_n = \det S_n$. The operator S belongs to the algebra $\mathbb{K}_N$ and its symbol coincides with the matrix-function $s(t)$. It follows from theorem 1.1 that

(3.13)
$$\lim_{n \to \infty} \frac{\det S_n}{G(s)^{n+1}} = \det(T \, (s^{-1}) \, S).$$

There remains only to show that $G(s)$ coincides with the right part of (3.12). In order to do so we write the equality

(3.14)
$$\begin{bmatrix} f & c \\ b & a_{mm} \end{bmatrix} = \begin{bmatrix} f & 0 \\ b & E_N \end{bmatrix} \begin{bmatrix} E_{N(m-1)} & f^{-1} c \\ 0 & a_{mm} - b \, f^{-1} \, c \end{bmatrix} ,$$

where E_k denotes the $\kappa \times \kappa$ identity matrix.

It follows from (3.14) that

$$\det(\text{symb } A)(t) = \det(\text{symb } F)(t) \; \det s(t)$$

and, hence,

$$G(s) = \exp \frac{1}{2\pi} \int_{\mathbb{T}} \log \frac{\det(\text{symb } A)(t)}{\det(\text{symb } F)(t)} \, |dt|.$$

The theorem is proved.

Theorem 3.4. *Let the conditions of theorem 3.3 be fulfilled. If operator A is invertible in $\ell_2^m(\mathbb{C}^N)$, then $\det(T(s^{-1}) \, S) \neq 0$, $\tilde{D}_n(A) \neq 0$ for all sufficiently large n and*

(3.15)
$$\lim_{n \to \infty} \frac{\tilde{D}_{n+1}(A)}{\tilde{D}_n(A)} = \exp \frac{1}{2\pi} \int_{\mathbb{T}} \log \frac{\det(\text{symb } A)(t)}{\det(\text{symb } F)(t)} \, |dt|.$$

In the proof of the theorem 3.3 we did not use the fact that all the operators A_{jk} belong to the algebra $\mathbb{K}_N$. We used only the fact that the operator $S = A_{mm} - B \, F^{-1} \, C$ belongs to $\mathbb{K}_N$.

Note that all the statements of this paper, without any changes in their proofs, *hold* if we replace M by a larger Krein algebra

$$\tilde{M} = \left\{ a \in L_\infty(\mathbb{T}) : \sum_{m \in \mathbb{Z}} |m \, a_m^2| < \infty \right\}$$

of not necessarily continuous functions. In this case, we replace $\mathbb{K}_N$ by the algebra $\tilde{\mathbb{K}}_N = \{T(a) + K\}$, where $a \in \tilde{M}^{N \times N}$ and the operators K are from the trace class.

In the case where the function $a(\in \tilde{M}^{N \times N})$ is not continuous, the integral

$$\int_{\mathbb{T}} \log \det a(t)|dt|$$

in the equality (1.6) has to be understood in the following sense

$$\lim_{r \uparrow 1} \int_{\mathbb{T}} \log \det(h_r a)(t)|dt|,$$

where $h_r a$ denotes the harmonic extension of a (for details se [BS, 1.36; 10.4]).

Let us finally remark that in this paper, by canonical factorization of $N \times N$ matrix functions, we always understand the canonical generalized factorization in the space $L_2^N(\mathbb{T})$ (for the definition see, for example, [CG, Ch. VII, §1]),

§4. Illustrative examples.

In this section we propose some illustrative examples.

1. Let $a, b, c, d, \in M$

$$(4.1) \qquad A = \begin{bmatrix} T(a) & T(b) \\ T(c) & T(d) \end{bmatrix} \quad \text{and} \quad R_n = \begin{bmatrix} I & 0 \\ 0 & P_n \end{bmatrix},$$

where $P_n \{\xi_k\}_{k=0}^{\infty} = (\xi_0, \xi_1, \ldots, \xi_n, 0, \ldots)$.

The following statement follows from theorem 3.3.

Theorem 4.1. *Let the operators $T(a)$ and $T(d - ba^{-1}c)$ be invertible in ℓ_2. Then*

$$(4.2) \qquad \lim_{n \to \infty} \frac{\tilde{D}_n(A)}{G(d - ba^{-1}c)^{n+1}} = \det\left(T\left(\frac{a}{ad - bc}\right)(T(d) - T(c)\, T^{-1}(a)\, T(b))\right),$$

where operator A is defined by (4.1) and the determinants $\tilde{D}_n(A)$ are defined by (3.4).

Moreover, if operator A is invertible in ℓ^2, then $\tilde{D}_n(A) \neq 0$ for all sufficiently large n and

$$(4.3) \qquad \lim_{n \to \infty} \frac{\tilde{D}_{n+1}(A)}{\tilde{D}_n(A)} = \exp \frac{1}{2\pi} \int_{\mathbb{T}} \log\left(\frac{ad - bc}{a}\right)|dt|.$$

2. Let $a, b, g, h \in M^{N \times N}$. Consider the operator

$$(4.4) \qquad A\xi = \begin{bmatrix} \cdot & \cdot & \cdot & \cdot & & \cdot & & \cdot & \cdot & \cdot \\ \cdot & \cdot & \cdot & a_0 & a_{-1} & g_{-2} & g_{-3} & \cdot & \cdot & \cdot \\ \cdot & \cdot & \cdot & a_1 & a_0 & g_{-1} & g_{-2} & \cdot & \cdot & \cdot \\ & & & & & & & & & \\ \cdot & \cdot & \cdot & h_2 & h_1 & b_0 & b_{-1} & \cdot & \cdot & \cdot \\ \cdot & \cdot & \cdot & h_3 & h_2 & b_1 & b_0 & \cdot & \cdot & \cdot \\ \cdot & \cdot & \cdot & & & & & \cdot & \cdot & \cdot \end{bmatrix} \begin{bmatrix} \cdot \\ \xi_{-2} \\ \xi_{-1} \\ \xi_0 \\ \xi_1 \\ \cdot \end{bmatrix}$$

acting in the space $\tilde{\ell}_2(\mathbf{C}^N)$ of two-sided sequences $\{\xi_k\}_{k=-\infty}^{\infty}$ $(\xi_k \in (\mathbf{C}^N)$. If $a = g$ and $b = h$ then A is a block paired Toeplitz operator; if $h = a$ and $g = b$. then A is transposed to such an operator; if $a = b = g = h$ then A is a Laurent operator.

Denote by $\{\tilde{R}_n\}$ the following sequence of projections

$$\tilde{R}_n\{\xi_k\}_{k=-\infty}^{\infty} = (\ldots \xi_{-2}, \xi_{-1}, \xi_0, \xi_1, \ldots, \xi_n, 0, 0 \ldots).$$

We suppose that $\det a(t) \neq 0$ and that the matrix a admits a canonical factorization: $a(t) = a_+(t) \, a_-(t)$. It follows from this that both operators $T(\tilde{a})$ and $T(a^{-1})$ are invertible in $\ell_2(\mathbf{C}^N)$.

Consider the sequence of determinants

(4.5) $d_n(A) = \det(\tilde{F}^{-1} \, \tilde{A}_n),$

where $\tilde{A}_n = \tilde{R}_n \, A \, \tilde{R}_n + I - \tilde{R}_n$ and $\tilde{F} = \tilde{A}_{-1}$. The invertibility of the operator $\tilde{F}^{-1}$ follows from the invertibility of $T(\tilde{a})$ (see (3.10) below).

Theorem 4.2. *Let* $a, b, g, h \in M^{N \times N}$. *Suppose that* $\det a(t) \neq 0$, $\det b(t) \neq 0$ $(t \in \mathbf{T})$ *and the matrices* a *and* b *admit canonical factorizations:* $a = a_+ \, a_-$, $b = b_+ \, b_-$. *Then for operator* A, *defined by (4.4), the following relation holds:*

(4.6) $\displaystyle \lim_{n \to \infty} \frac{d_n(A)}{G(b)^{n+1}} = \det(T(b^{-1}) \, (T(b) - H(h) \, T^{-1}(\tilde{a}) \, H(\tilde{g})).$

In particular, if $g = a := \alpha$, $h = b := \beta$, *then*

(4.7) $\displaystyle \lim_{n \to \infty} \frac{d_n(A)}{G(\beta)^{n+1}} = \det(T(\alpha_- \beta^{-1}) \, T(\beta \alpha_-^{-1}));$

If $h = a := \gamma$ *and* $g = b := \delta$, *then*

(4.8) $\displaystyle \lim_{n \to \infty} \frac{d_n(A)}{G(\delta)^{n+1}} = \det(T(\gamma_+ \delta^{-1}) \, T(\delta \gamma_+^{-1}));$

If $a = b = c = d$, *then*

(4.9) $\displaystyle \lim_{n \to \infty} \frac{d_n(A)}{G(a)^{n+1}} = 1.$

Proof. In order to use theorem 3.3, we transform the space $\tilde{\ell}_2(\mathbf{C}^N)$ into $\ell_2^2(\mathbf{C}^N)$ by isomorphism

$$\pi\{\xi_k\}_{k=-\infty}^{\infty} = (\{\xi_{-k-1}\}_{k=0}^{\infty}, \; \{\xi_k\}_{k=0}^{\infty}),$$

and we obtain

$$(4.10) \qquad \pi \, A \pi^{-1} = \begin{bmatrix} T(\tilde{a}) & H(\tilde{g}) \\ H(h) & T(b) \end{bmatrix}, \quad \pi \, \tilde{R}_n \, \pi^{-1} = \begin{bmatrix} I & 0 \\ 0 & P_n \end{bmatrix}.$$

Let $B = \pi \, A \, \pi^{-1}$ and $R_n = \pi \, \tilde{R}_n \, \pi^{-1}$, then the determinant $d_n(A)$ coincides with $\tilde{D}_n(B)$, defined by (3.4).

Denote by S the operator

$$(4.11) \qquad S := T(b) - H(h) \, T^{-1}(\tilde{a}) \, H(\tilde{g}).$$

Since $H(h)$ and $H(\tilde{g})$ are compact operators, symb $S = b$. By theorem 3.3

$$(4.12) \qquad \lim_{n \to \infty} \frac{d_n(A)}{G(b)^{n+1}} = \det(T(b^{-1}) \, (T(b) - H(h) \, T^{-1}(\tilde{a}) \, H(\tilde{g}))).$$

Using the well-known relations (see for example [BS; 2.14])

$$T(ab) = T(a) \, T(b) + H(a) \, H(\tilde{b})$$

$$H(ab) = T(a) \, H(b) + H(a) \, T(\tilde{b})$$

$$H(h) \, T(\tilde{a}_-^{-1}) = H(ha_-^{-1}), \quad T(\tilde{a}_+^{-1}) \, H(\tilde{g}) = H\left(\widetilde{a_+^{-1}g}\right)$$

the equality (4.12) can be rewritten in the form

$$(4.13) \qquad \lim_{n \to \infty} \frac{d_n(A)}{G(b)^{n+1}} = \det(T(b^{-1}) \, (T(b - ha^{-1}g) + T(ha_-^{-1}) \, T(a_+^{-1}g))).$$

If, in particular, $g = a, h = b$ then the right part of (4.13) equals

$$\det T(b^{-1}) \, T(ba_-^{-1}) \, T(a_-) = \det T(a_-) \, T(b^{-1}) \, T(ba_-^{-1}) = \det T(a_- b^{-1}) \, T(ba_-^{-1})$$

and, hence, (4.7) is valid.

The relation (4.8) can be checked analogously. Finally, (4.9) follows from (4.6).

Corollary 4.3. *Let the conditions of theorem 4.2 be fulfilled and, in addition, operator A defined by (4.4) is invertible in $\tilde{\ell}_2(\mathbb{C}^N)$. Then $d_n(A) \neq 0$ for all sufficiently large n and*

$$(4.14) \qquad \lim_{n \to \infty} \frac{d_{n+1}(A)}{d_n(A)} = \exp \frac{1}{2\pi} \int_{\mathbb{T}} \log \det b(t) \, |dt|.$$

3. Let A again denote the operator, defined by the equality (4.4). We consider here Szegö-Widom type limit theorems for this operator, relative to another sequence of projections. Consider the following sequence of matrices

$$A_n = \begin{bmatrix} a_0 & \cdots & a_{-n} & g_{-n-1} & \cdots & g_{-2n-1} \\ \cdot & \cdots & \cdot & \cdot & \cdots & \cdot \\ a_n & \cdots & a_0 & g_{-1} & \cdots & g_{-n-1} \\ h_{n+1} & \cdots & h_1 & b_0 & \cdots & b_{-n} \\ \cdot & \cdots & \cdot & \cdot & \cdots & \cdot \\ h_{2n+1} & \cdots & h_{n+1} & b_n & \cdots & b_0 \end{bmatrix}$$

Theorem 4.4. Let $a, b, g, h \in C(\mathbb{T})^{N \times N}$. Suppose that the Toeplitz operators $T(\tilde{a})$ and $T(b)$ are invertible in $\ell_2(\mathbb{C}^N)$ and the Hankel operator $H \begin{bmatrix} 0 & \tilde{g} \\ h & 0 \end{bmatrix}$ is of the trace class in the space $\ell_2(\mathbb{C}^{2N})$. Then

(4.15)
$$\lim_{n \to \infty} \frac{\det A_n}{G(\tilde{a}b)^{n+1}}$$

$$= \det \left(T(\tilde{a}^{-1}) \, T(\tilde{a}) \right) \det \left(T(b^{-1}) \, T(b) \right) \det \left(I + T^{-1} \begin{bmatrix} \tilde{a} & 0 \\ 0 & b \end{bmatrix} H \begin{bmatrix} 0 & \tilde{g} \\ h & 0 \end{bmatrix} \right).$$

If, moreover, the operator A is invertible in $\tilde{\ell}_2(\mathbb{C}^N)$, then

(4.16)
$$\lim_{n \to \infty} \frac{\det A_{n+1}}{\det A_n} = G(\tilde{a}b).$$

Proof. Consider the following sequence of projections acting in $\tilde{\ell}_2(\mathbb{C}^N)$:

$$\tilde{P}_n \{\xi_k\}_{k=-\infty}^{\infty} = (\ldots 0, 0, \xi_{-n-1}, \ldots, \xi_{-1}, \xi_0, \ldots, \xi_n, 0, 0, \ldots)$$

It is readily checked that

$$\pi A \pi^{-1} = \begin{bmatrix} T(\tilde{a}) & H(\tilde{g}) \\ H(h) & T(b) \end{bmatrix} , \pi \tilde{P}_n \pi^{-1} = \begin{bmatrix} P_n & 0 \\ 0 & P_n \end{bmatrix}$$

(compare with (4.10)). Operator $\pi A \pi^{-1}$ can be rewritten in the form $T(x) + H(y)$, where

$$x = \begin{bmatrix} \tilde{a} & 0 \\ 0 & b \end{bmatrix} , \; y = \begin{bmatrix} 0 & \tilde{g} \\ h & 0 \end{bmatrix} \quad (\in \mathbb{C}^{2N \times 2N}).$$

The operators $B := T(x) + H(y)$ and $\mathbb{P}_n := \text{diag}\,(P_n, P_n)$ act in the space $\ell_2(\mathbb{C}^{2N})$, $B \in \mathbb{K}_{2N}$, symb $B = x$ and $\det(T(x^{-1})B)$ coincides with the right part of the equality (4.15). Thus, all the statements in Theorem 4.4 follow from Theorems 1.1 and 1.2.

Note that the equalities (4.15), (4.16) represent discrete analogues of Szegö formulas for a class of generalized Toeplitz integral operators obtained in [D].

§5. Proofs of theorems 2.1, 2.2.

Let A be the operator defined by (2.2) and π the isomorphism

$$\pi : \ell_2(\mathbb{C}^N) \times \tilde{\ell}_2(\mathbb{C}^N) \to \ell_2^3(\mathbb{C}^N)$$

defined by

$$\pi \left(\{\eta_k\}_{k=0}^{\infty}, \; \{\xi_k\}_{k=-\infty}^{\infty} \right) = \left(\{\eta_k\}_{k=0}^{\infty}, \; \{\xi_{-k-1}\}_{k=0}^{\infty}, \; \{\xi_k\}_{k=0}^{\infty} \right).$$

It is readily checked that

$$(5.1) \qquad \pi \, A \, \pi^{-1} = \begin{bmatrix} T(u) & H(g) & T(g) \\ H(\tilde{h}) & T(\tilde{v}) & H(\tilde{v}) \\ T(h) & H(v) & T(v) \end{bmatrix},$$

$$(5.2) \qquad \pi \, \tilde{R}_n \, \pi^{-1} = \begin{bmatrix} I_N & 0 & 0 \\ 0 & I_N & 0 \\ 0 & 0 & P_m \end{bmatrix} \qquad (m = -n \geq 0)$$

and

$$(5.3) \qquad \pi \, \tilde{R}_n \, \pi^{-1} = \begin{bmatrix} I_N & 0 & 0 \\ 0 & Q_N & 0 \\ 0 & 0 & 0 \end{bmatrix} \qquad (n \geq 0)$$

where $P_n\{\xi_k\}_0^\infty = (\xi_0, \ldots, \xi_n, 0, \ldots)$ and $Q_n = I_N - P_n$.

We consider separately the cases $n \to -\infty$ and $n \to +\infty$. We start with the case $n < 0$. In this case, $d(A_n) = \tilde{D}_n(B)$, where $B = \pi \, A \, \pi^{-1}$ and the determinants $d(A_n)$ and $\tilde{D}_n$ are defined respectively by (2.4) and (3.4).

It is clear that

$$(5.4) \qquad \text{symb } B = \begin{bmatrix} u & 0 & g \\ 0 & \tilde{v} & 0 \\ h & 0 & v \end{bmatrix}, \quad \text{symb } F = \begin{bmatrix} u & 0 \\ 0 & \tilde{v} \end{bmatrix}$$

and

$$(5.5) \qquad \frac{\det \text{symb } B}{\det \text{symb } F} = \det(v - hu^{-1}g) \quad (= \det w).$$

Using the equalities (5.1) - (5.5) it is evident that all the statements of theorem 2.1 follow directly from Theorems 3.3 and 3.4.

Let now $n > 0$ ($n \to +\infty$). In this case, $d(A_n) = \det(F^{-1} \, \tilde{B}_n)$, where

$$(5.6) \qquad F = \begin{bmatrix} T(u) & H(g) \\ H(\tilde{h}) & T(\tilde{v}) \end{bmatrix} \quad \text{and} \quad \tilde{B}_n = \begin{bmatrix} T(u) & H(g) \, Q_n \\ Q_n \, H(\tilde{h}) & Q_n \, T(\tilde{v}) \, Q_n + P_n \end{bmatrix}.$$

Under the conditions of theorem 2.3, the operators F and $T(u)$ are invertible in the space $\ell_2^2(\mathbb{C}^N)$ and $\ell_2(\mathbb{C}^N)$ respectively. In this case,

$$(5.7) \qquad F^{-1} = \begin{bmatrix} \Delta_1^{-1} & -T^{-1}(u) \, H(g) \, \Delta^{-1} \\ -\Delta^{-1} \, H(\tilde{h}) \, T^{-1}(u) & \Delta^{-1} \end{bmatrix},$$

where

$$\Delta = T(\tilde{v}) - H(\tilde{h}) \, T^{-1}(u) \, H(g)$$

and

$$\Delta_1 = T(u) - H(g) \, T^{-1}(\tilde{v}) \, H(\tilde{h}).$$

Note that in the decomposition of the space $\ell_2^2(\mathbb{C}^N)$ into the product $\ell_2(\mathbb{C}^N) \times Q_n \ell_2(\mathbb{C}^N) \times P_n \ell_2(\mathbb{C}^N)$ the operators $F, \tilde{B}_n$ and F^{-1} have the following block representations:

$$(5.8) \qquad F = \begin{bmatrix} T(u) & K_n & * \\ L_n & S_n & * \\ * & * & * \end{bmatrix}, \quad \tilde{B}_n = \begin{bmatrix} T(u) & K_n & 0 \\ L_n & S_n & 0 \\ 0 & 0 & P_n \end{bmatrix},$$

and

$$(5.9) \qquad F^{-1} = \begin{bmatrix} * & * & * \\ * & * & * \\ * & * & P_n \Delta^{-1} P_n \end{bmatrix},$$

where, with stars, we denote the entries, the explicit form of which does not matter. Note that P_n is a finite-dimensional projection. Using the Schur decomposition (3.10) for the matrix $F^{-1}B_n$ and representation (5.9), we obtain the relation

$$d(B_n) = \det(B_0^{-1} \, \tilde{B}_n) = \det P_n \, \Delta^{-1} \, P_n.$$

The operator Δ belongs to the algebra $\mathbb{K}_N$ considered in section 1, and symb $\Delta = \tilde{v}$. Thus, all the statements in theorem 2.2 follow from theorems 1.1 and 1.2.

REFERENCES

[AAK] Adamjan, V.M., Arov, D.Z., Krein, M.G., *Infinite Hankel block matrices and related extension problems*, Izv. Akad. Nauk Armjan, SSR Ser. Math. (Russian) **6** (1971), 87-112; Amer. Math. Soc. Transl. (2) **111** (1978), 133-156.

[BS] Böttcher, A., Silbermann, B., *Analysis of Toeplitz Operators*, Akademie - Verlag, Berlin, 1989.

[CG] Clancey, K., Gohberg, I., *Factorization of Matrix Functions and Singular Integral Operators*, OT3, Birkhäuser - Verlag, Basel, 1981.

[D] Dym, H., *On a Szegö formula for a class of generalized Toeplitz kernels*, Integr. Equat. and Operator Theory **8** (1985), 427-431.

[DF] Dym, H., Gohberg, I., *A maximum entropy principle for contractive interpolants*, J. of Functional Analysis **65** (1986), 83-125.

[EG] Ellis, R.L., Gohberg, I., *Orthogonal systems related to infinite Hankel matrices*, J. of Functional Analysis **109. N. 1** ((1992)), 155-198.

[G] Gohberg, I.C., *On an application of the theory of normed rings to singular integral equations*, Uspekhi mat. nauk **72** (1952), 149-156.

[K] Krein, M.G., *On some new Banach algebras and theorems of Wiener-Levy type for Fourier series and integrals*, (Russian), Mat. Issled. **1:1** (1966), 82-109; (Eng. transl), Amer. Math. Soc. Transl. **932** (1970), 177-199.

[W1]   Widom, H., *Perturbing Fredholm operators to obtain invertible operators*, J. Funct. Anal **29** (1975), 26-31.

[W2]   _____, *Asymptotic behavior of block Toepelitz matrices and determinats, II*, Adv. Math. **21** (1976), 1-29.

I. Gohberg
School of Mathematical Sciences
Raymond and Beverly Sackler Faculty of Exact Sciences
Tel-Aviv University
69978 Tel-Aviv
Israel

N. Krupnik
Department of Mathematics and Computer Science
Bar-Ilan University
52900 Ramat-Gan
Israel

MSC 1991: 47B53

Operator Theory:
Advances and Applications, Vol. 71
© 1994 Birkhäuser Verlag Basel/Switzerland

(SEMI)-FREDHOLMNESS OF CONVOLUTION OPERATORS ON THE SPACES OF BESSEL POTENTIALS

Yuri Karlovich and Ilya Spitkovsky [1]

Dedicated to Harold Widom on the occasion of his 60th birthday

The consideration of above mentioned operators on the union of intervals and/or rays is reduced to the canonical situation of operators W_K on $L_p(\mathbb{R}_+)$ with semi almost periodic presymbols K at the expense of inflating the size of K. The Fredholm theory (that is, conditions of n-, d-normality and the index formula) is developed. In particular, relations between (semi-)Fredholmness of W_K, invertibility of $W_{K_\pm}$ with $K_\pm$ being almost periodic representatives of K at $\pm\infty$, and factorability of $K_\pm$ are established.

1. Introduction.

Let $S'(\mathbb{R})$ be the space of all temperate distributions on $\mathbb{R}$, $\mathcal{F}$ and $\mathcal{F}^{-1}$ — correspondingly the direct and inverse Fourier transforms, acting on $S'(\mathbb{R})$ and defined on infinitely differentiable functions ψ with compact support in $\mathbb{R}$ according to the formula:

$$\left(\mathcal{F}^{\pm 1}\psi\right)(x) = \frac{1}{\sqrt{2\pi}} \int_{\mathbb{R}} e^{\mp ixy} \psi(y)\, dy.$$

Let us denote by $\mathcal{H}_{s,p}$ $(1 < p < \infty, s \in \mathbb{R})$ the space of Bessel potentials, which can be defined as follows:

(1.1)
$$\mathcal{H}_{s,p}(\mathbb{R}) = \{f \in S'(\mathbb{R}) : \varphi = \mathcal{F}^{-1}(1 + x^2)^{s/2}\mathcal{F}f \in L_p\}.$$

[1] Partially supported by NSF Grant DMS-91-01143

Obviously, $\mathcal{H}_{0,p}(\mathbb{R}) = L_p(\mathbb{R})$. The norm on $\mathcal{H}_{s,p}(\mathbb{R})$ is given by $||f||_{s,p} = ||\varphi||_p$. Let Ω be a finite union of rays and intervals of $\mathbb{R}$ separated from each other. Let us denote by $\overset{0}{\mathcal{H}}_{s,p}(\Omega)$ the set of all functions from $\mathcal{H}_{s,p}(\mathbb{R})$ having their support in $\overline{\Omega}$, and by $\mathcal{H}_{s,p}(\Omega)$ the set of restrictions onto Ω of functions from $\mathcal{H}_{s,p}(\mathbb{R})$; r_Ω stands for the corresponding restriction operator. The norm on $\overset{0}{\mathcal{H}}_{s,p}(\Omega)$ is induced from $\mathcal{H}_{s,p}(\mathbb{R})$, and $\mathcal{H}_{s,p}(\Omega)$ is supplied with the norm of the factor-space $\mathcal{H}_{s,p}(\mathbb{R})/\overset{0}{\mathcal{H}}_{s,p}(\mathbb{R}\setminus\overline{\Omega})$. It is well known that the set $C_0^\infty(\Omega)$ of infinitely differentiable functions with compact in Ω support is dense in $\overset{0}{\mathcal{H}}_{s,p}(\Omega)$ (see, for example, [36]).

For any $k \in S'(\mathbb{R})$ we will denote by W_K^0 the corresponding convolution operator: $W_K^0 u = k * u = \mathcal{F}^{-1}K\mathcal{F}u$, where $K = \mathcal{F}k$.

This paper is devoted to the study of equations of the form

$$(1.2) \qquad r_\Omega(k * u) = f,$$

where unknown u and given f are n-dimensional vector-functions with elements in $\overset{0}{\mathcal{H}}_{\sigma,p}(\Omega)$ and $\mathcal{H}_{s,p}(\Omega)$ correspondingly, and k is a given $n \times n$ matrix function with its elements in $S'(\mathbb{R})$. Equation (1.2) can be rewritten as $W_{K,\Omega}u = f$, where

$$(1.3) \qquad W_{K,\Omega} = r_\Omega W_K^0 : \overset{0}{\mathcal{H}}_{\sigma,p}(\Omega) \to \mathcal{H}_{s,p}(\Omega).$$

We are interested in the *Fredholm theory* of operators $X = W_{K,\Omega}$, that is, in necessary and sufficient conditions for these operators to be either *d-normal* (i.e., have closed range Im X of finite codimension $\beta(X)$), *n-normal* (i.e., have closed range and a kernel of finite dimension $\alpha(X)$), or *Fredholm* (*d*-normal **and** *n*-normal). The operator X is called *semi-Fredholm* if it is *n*-normal **or** *d*-normal. Let us remind that $\alpha(X)$ and $\beta(X)$ are called the *defect numbers* of the operator X, their difference ind $X = \alpha(X) - \beta(X)$ is called the *index* of X, and it is finite for Fredholm operators. The calculation of ind X in the latter case is also a part of Fredholm theory for X.

We will consider only bounded operators. According to the definiton (1.1), a sufficient condition for boundedness of (1.3) is given by

$$(x^2 + 1)^{-\mu/2} K(x) \in M_p \text{ for some } \mu \leq \sigma - s,$$

where M_p $(1 < p < \infty)$ is the set of all functions $a \in L_\infty(\mathbb{R})$ such that the operator $\mathcal{F}^{-1}a\mathcal{F}$ is bounded on $L_p(\mathbb{R})$.

To review known results on Fredholm properties of $W_{K,\Omega}$, as well as to explain the purpose of this paper, we have to introduce some subclasses of M_p.

It is well known [12] that $\mathcal{M}_p$ is a Banach algebra with a norm $||a||_p = ||W_a^0||_{L_p \to L_p}$. We will denote by $\mathcal{M}_p^0$ the closure in $\mathcal{M}_p$ of the set $\bigcup_{r \in R_p} \mathcal{M}_r$, where

(1.4) $$R_p = \{x: |\frac{1}{x} - \frac{1}{2}| > |\frac{1}{p} - \frac{1}{2}|\}.$$

As usual, $C(\overline{\mathbb{R}})$ and $C(\dot{\mathbb{R}})$ will stand correspondingly for the space of continuous functions on the two point compactification $\overline{\mathbb{R}}$ and one point compactification $\dot{\mathbb{R}}$ of the real line $\mathbb{R}$.

Denote by AP_W the set of all almost periodic functions with absolutely convergent Fourier series:

$$APW = \{f = \sum a_\nu e^{i\nu x} : \sum |a_\nu| < \infty\}$$

with the norm $||f||_{APW} = \sum |a_\nu|$. Since $\mathcal{F}^{-1} e^{i\nu x} \mathcal{F}$ is nothing but a shift by ν, for all $p \in (1, \infty)$ $APW \subset \mathcal{M}_p$ and $||f||_{APW} \geq ||f||_p$. Let AP_p be the closure of APW in $\mathcal{M}_p$. Finally, put $\Pi C_p^\infty(\mathbb{R}) = C(\overline{\mathbb{R}}) \cap \mathcal{M}_p^0$.

Many papers have been devoted to the canonical case $s = \sigma = 0$, $\Omega = \mathbb{R}_+$, where

(1.5) $$W_{K, \mathbb{R}_+} (\overset{\text{def}}{=} W_K) : L_p(\mathbb{R}_+) \to L_p(\mathbb{R}_+).$$

The Fredholm theory of such operators with presymbols K in the Wiener algebra $\mathbb{C} \dot{+} \mathcal{F}L_1$ ($\subset \mathcal{M}_p$) was developed in [11]. The corresponding results for $K \in \Pi C_p^\infty(\mathbb{R})$ were obtained in [7]. Scalar ($n = 1$) operators W_K with $K \in AP_p$ were dealt with in [9, 10]. Finally, the algebra SAP_p generated by AP_p and ΠC_p^∞ (also in a scalar case) was a subject of consideration in [8].

It is worth mentioning that the Banach algebra $\mathcal{M}_2$ coincides with L_∞, and therefore $\Pi C_2^\infty(\mathbb{R}) = C(\overline{\mathbb{R}})$ and $AP_2 (= AP)$ is a usual set of all Bohr almost periodic functions. The corresponding algebra $SAP_2 (= SAP)$ generated by AP and $C(\overline{\mathbb{R}})$ was introduced in [28].

Since the Fourier transform $\mathcal{F}$ implements an endomorphism of $L_2(\mathbb{R})$, and $\mathcal{F}r_{\mathbb{R}_\mp} \mathcal{F}^{-1} = P_\pm = \frac{1}{2}(I \pm S)$, where S is a singular integral operator with Cauchy kernel:

(1.6) $$(S\varphi)(t) = \frac{1}{\pi i} \int_{-\infty}^{+\infty} \frac{\varphi(\tau)}{\tau - t} d\tau,$$

the operator W_K on $L_2(\mathbb{R}_+)$ has the same Fredholm properties as the Toeplitz operator $\mathcal{F}W_K \mathcal{F}^{-1} = P_- K \mid \text{Im } P_-$, or the singular integral operator

$$T_K = P_+ + KP_- : L_2(\mathbb{R}) \to L_2(\mathbb{R}).$$

Therefore, the Fredholm theory for the convolution type operators W_K in $L_2(\mathbb{R}_+)$ is a reformulation of the corresponding earlier results for operators T_K. A necessary condition for T_K to be semi-Fredholm is invertibility of its symbol K: $K^{-1} \in L_\infty$. If $K \in C(\dot{\mathbb{R}})$ or $K \in AP$, $n = 1$, this condition is also sufficient, all semi-Fredholm operators T_K with continuous symbols K are Fredholm, and these results do not depend on p (see [4, 22] for references in the case $K \in C(\dot{\mathbb{R}})$, and [9, 5] for scalar AP symbols). For the class $C(\overline{\mathbb{R}})$ though Fredholmness of T_K depends on the relation between p and jump of K at ∞ (see [4, 22]). H. Widom [37] was among those who discovered and analised this dependence, which, after an appropriate generalization of a "jump" concept, extends to the SAP setting [28, 27]. It is our pleasure to publish this paper in a special issue devoted to H. Widom.

The case of a finite interval $\Omega = (0, \lambda)$, $n = 1$, $p = 2$, and nonzero σ, s such that $(x^2 + 1)^{-\alpha/2} K \in C(\overline{\mathbb{R}})$, $\alpha = \sigma - s$ was considered in [24, 25]. Our paper [16] was devoted to both infinite ($\Omega = \mathbb{R}_+$) and finite ($\Omega = (0, \lambda)$) cases, $p = 2$, arbitrary n, and s, σ satisfying

$$(x^2 + 1)^{-\alpha/2} K \in SAP \text{ for some } \alpha \in \mathbb{C}, \operatorname{Re} \alpha = \sigma - s.$$

We promised in [16] to return to the case $p \neq 2$ in future publications. For various reasons (such as emigration of one of the authors from the Soviet Union in 1989) it was difficult for us to fulfil this promise until we met again in Germany in summer of 1992, and discussed there some details of this paper. We would like to thank Prof. E. Meister from Darmstadt and Prof. B. Silbermann from Chemnitz whose professional invitations made this meeting possible.

2. Main results.

Denote by $c_0, \ldots, c_{m-1}$ all the endpoints (in the increasing order) of intervals forming Ω. Put also $c_{-1} = -\infty$, $c_m = +\infty$, $\Omega_j = (c_{j-1}, c_j)$ $(j = 0, \ldots, m)$, and $\lambda_j = c_j - c_{j-1}$ $(j = 1, \ldots, m-1)$.

Suppose that for some $\alpha \in \mathbb{C}$

$$(2.1) \qquad N(x) \overset{\text{def}}{=} (x^2 + 1)^{-\alpha/2} K(x) \in SAP_p.$$

Denoting $J = \{j \in \{0, \ldots, m\}: \Omega_j \subset \Omega\}$, and $\overline{J} = \{0, \ldots, m\} \setminus J$, let us put

$$(2.2) \qquad N_j(x) = \begin{cases} N(x) & \text{for } j \in J \\ I & \text{for } j \in \overline{J} \end{cases}$$

and

$$(2.3) \qquad \alpha_j = \begin{cases} \alpha, & j \in J \\ 0, & j \in \overline{J} \end{cases}, \quad s_j = \begin{cases} \sigma, & j \in J \\ s, & j \in \overline{J} \end{cases}.$$

THEOREM 2.1. *Let in (2.1) $\operatorname{Re} \alpha \leq \sigma - s$. Then the (bounded) operator $W_{K,\Omega}$ given by (1.3) is Fredholm, n-normal, d-normal if and only if*

$$(2.4) \qquad N^{-1} \in SAP_p \text{ (in case of unbounded } \Omega), \quad \operatorname{Re} \alpha = \sigma - s,$$

and the operator $W_{\tilde{K}} : L_p(\mathbb{R}_+) \to L_p(\mathbb{R}_+)$ has the above mentioned property. Moreover, the corresponding defect numbers of $W_{K,\Omega}$ and $W_{\tilde{K}}$ coincide.

Here $\tilde{K}$ is an $nm \times nm$ matrix function given by the formula

$$(2.5) \qquad \tilde{K} = \begin{bmatrix} f_1 & & & 0 \\ & \ddots & & \vdots \\ & & f_{m-1} & 0 \\ g_1 & \cdots & g_{m-1} & g_m \end{bmatrix},$$

with

$$(2.6) \qquad \begin{aligned} f_j(x) &= \left(\frac{x-i}{x+i}\right)^{-s_j - i \operatorname{Im} \alpha_j} e^{i\lambda_j x} I_n && (j = 1, \ldots, m-1) \\ g_j(x) &= \left(\frac{x-i}{x+i}\right)^{-s - \frac{a_0 + a_j}{2}} e^{-i(c_{j-1} - c_0)x} N_0^{-1}(x) N_j(x) && (j = 1, \ldots, m), \end{aligned}$$

and all the elements of (2.5) outside the main diagonal or last row are equal zero. Of course, $\tilde{K}^{\pm 1} \in SAP_p$ whenever the conditions (2.1) and (2.4) are met.

Theorem 2.1 shows that the case of general Ω and arbitrary s, σ can be reduced to the canonical case $s = \sigma = 0$, $\Omega = \mathbb{R}_+$ at the expense of "m-ling" of the problem size (passing from K to $\tilde{K}$ given by (2.5)). It is crucial that the blocks f_j of $\tilde{K}$ contain elementary almost periodic multiples $e^{i\lambda_j x}$, which makes consideration of matrix SAP_p presymbols unavoidable (even for scalar piecewise continuous K) if one wishes to cover the case $m > 1$, when Ω differs from the half-line.

Due to Theorem 2.1, we may concentrate mostly on operators (1.5) with $K \in SAP_p$ $(\subset SAP)$. According to [28], for any such K there exist the unique matrix functions $K_\pm \in AP$ such that

$$\lim_{x \to \pm\infty} (K(x) - K_\pm(x)) = 0,$$

and $K \mapsto K_\pm$ is an algebraic homomorphism of SAP onto AP. We will call $K_\pm$ the (almost periodic) *local representatives* of K in $\pm\infty$. It will be shown in Lemma 3.1 that $K_\pm \in AP_p$ whenever $K \in SAP_p$.

Condition (2.4) for the operator (1.5) means simply that $K^{-1} \in SAP_p$. Then $(K_\pm)^{-1} = (K^{-1})_\pm \in AP_p$, and therefore invertibility in AP_p of the local representatives $K_\pm$ of K $(\in SAP_p)$ is necessary for the operator (1.5) to be semi-Fredholm.

We presume (but were not able to prove in its full generality) that the following stronger result holds.

CONJECTURE. *Let the operator (1.5) with $K \in SAP_p$ be n- (d-) normal in $L_p(\mathbb{R}_+)$. Then the corresponding operators $W_{K_\pm}$ are left (right) invertible in $L_p(\mathbb{R}_+)$.*

This conjecture is valid for $n = 1$ (see [9, 10] for the case $K_+ = K_-$, and [8] in general scalar case). The next two theorems show that it is also true when $p = 2$, and for Fredholm operators.

THEOREM 2.2. *If the operator* (1.5) *with* $K \in SAP$ *is n- (d-)normal in* $L_2(\mathbb{R}_+)$, *then the corresponding operators* $W_{K_\pm}$ *are left (right) invertible in* $L_2(\mathbb{R}_+)$.

THEOREM 2.3. *If the operator* (1.5) *with* $K \in SAP_p$ *is Fredholm, then the corresponding operators* $W_{K_\pm}$ *are invertible.*

Therefore, our next concern is the invertibility (one- and two-sided) of operators

$$(2.7) \qquad\qquad W_G \colon L_p(\mathbb{R}_+) \to L_p(\mathbb{R}_+)$$

with $G \in AP_p$. This is closely related (though not as closely as we would wish) with the so called *almost periodic* (abbreviated: *AP*)-*factorization* of G. Recall that, according to [16], an AP-factorization of an $n \times n$ matrix function G is its representation in the form

$$(2.8) \qquad\qquad G(x) = G^+(x)\, \Lambda(x)\, G^-(x),$$

where $\Lambda(x) = \text{diag}[e^{i\lambda_1 x}, \dots, e^{i\lambda_n x}]$,

$$(2.9) \qquad\qquad (G^+)^{\pm 1} \in AP^+, \quad (G^-)^{\pm 1} \in AP^-,$$

and

$$AP^\pm = \{f \in AP \colon \mathbf{M}(e^{-i\lambda x} f(x)) = 0 \text{ for } \lambda \in \mathbb{R}_\mp \cup \{0\}\}.$$

Here and in what follows $\mathbf{M}(\varphi)$ stands for the *Bohr mean value* of $\varphi (\in AP)$:

$$\mathbf{M}(\varphi) = \lim_{T \to \infty} \frac{1}{2T} \int_{-T}^{T} \varphi(\tau)\, d\tau.$$

If the factorization (2.8) exists, then the *partial AP-indices* $\lambda_1, \dots, \lambda_n$ are defined by G uniquely, and, if all of them are the same, then the matrix

$$d(G) = \mathbf{M}(G^+)\,\mathbf{M}(G^-) \ (\in \mathbb{C}^{n \times n})$$

is also unique. See [16, Section 2] for these and other properties of AP-factorization.

We will say that (2.8) is an AP_p- (AP_W-) factorization of G if in (2.9) the classes $AP^\pm$ are substituted by $AP_p^\pm = AP_p \cap AP^\pm$ ($AP_W^\pm = AP_W \cap AP^\pm$). Of course, any AP_W-factorization is at the same time an AP_p-factorization for all $p \in (1, \infty)$.

The next result is almost obvious (and therefore will **not** be supplied with a proof in sections to follow).

THEOREM 2.4. *Let a matrix function* G *admit an* AP_p-*factorization* (2.8). *Then the operator* (2.7) *is n-normal (d-normal, Fredholm) if and only if all the AP-partial indices* in (2.8) are nonpositive (nonnegative, zero). If this condition is satisfied, then W_G is left (right, two-side) invertible.

Let us mention that, according to Theorem 2.3, all the Fredholm operators W_G with $G \in AP_p$ are invertible, independent of the AP_p-factorability of G.

All scalar functions G which are invertible in SAP_p generate one side invertible operators W_G [8], but not all such functions (even those corresponding to invertible W_G) admit an AP-factorization. Moreover, for any $n > 1$ there exist $n \times n$ matrices G with almost periodic polynomial entries and $\det G = \text{const} \neq 0$ with no AP-factorization at all [16]. When multiplied by $e^{i\lambda x}$ with a suitable λ, these G deliver examples of matrices from $AP_W^{\pm}$, invertible in AP_W and not AP-factorable. In spite of that, the corresponding operators W_G are left/right invertible (for all p). These examples mean that AP_p- (and even AP-) factorability condition is not necessary for semi-Fredholmness of W_G.

Nevertheless, it becomes necessary under certain circumstances.

THEOREM 2.5. If $G \in AP_W$ and the operator (2.7) is Fredholm, then G is AP_W-factorable.

In the next three theorems of this section we suppose that local representatives at $\pm\infty$ of presymbols involved are AP_p-factorable. According to Theorems 2.3 and 2.5, this assumption becomes redundant when Fredholmness is considered and $K_\pm \in AP_W$.

THEOREM 2.6. Let Ω, K and $\tilde{K}$ be as in Theorem 2.1, and let $\tilde{K}_\pm$ (the local representatives of $\tilde{K}$) admit AP_p-factorizations

$$(2.10) \qquad \tilde{K}_\pm(x) = \tilde{K}_\pm^+(x)\,\Lambda_\pm(x)\,\tilde{K}_\pm^-(x).$$

Then for the operator (1.3) to be Fredholm it is necessary and sufficient that:

(i)   (2.4) is satisfied,
(ii) all the AP-partial indices of $\tilde{K}_\pm$ equal zero,
(iii) for all the eigenvalues ξ_j $(j = 1, \ldots, mn)$ of the matrix $d(\tilde{K}_-)^{-1} d(\tilde{K}_+)$,

$$(2.11) \qquad \gamma_j \overset{\text{def}}{=} \{\frac{1}{q} + \frac{\sigma+s}{2} - \frac{1}{2\pi} \arg \xi_j\} \neq 0.$$

If conditions (i)–(iii) hold, then

$$(2.12) \qquad \text{ind}\, W_{K,\Omega} = \text{Ind}\, \det \tilde{K} - mn\left(\frac{1}{q} + \frac{\sigma+s}{2}\right) + \sum_{j=1}^{mn} \gamma_j$$

If conditions (i), (ii) are satisfied, and (iii) fails, then $W_{K,\Omega}$ has finite defect numbers $\alpha(W_{K,\Omega})$ and $\beta(W_{K,\Omega})$, but its range is not closed.

Here, as in [16], Ind is a functional defined according to the formula:

$$\text{Ind}\, f = \frac{1}{2\pi} \lim_{t\to+\infty} \frac{1}{t} \int_{-t}^{t} \arg f(s)\, \text{sign}\, s\, ds.$$

In case of semi-Fredholm (but not Fredholm) operator W_K not all the partial indices λ_j are equal 0. Let us denote by l^+ (l^-) the number of nonzero indices among $\lambda_1^+, \ldots, \lambda_{mn}^+$ $(\lambda_1^-, \ldots, \lambda_{mn}^-)$. It is convenient to choose the order of diagonal elements of $\Lambda_\pm$ in such a way that nonzero partial indices go first. Let $\Phi^+ = M(\tilde{K}_-^+)^{-1} M(\tilde{K}_+^+), \Phi^- = M(\tilde{K}_+^-) M(\tilde{K}_-^-)^{-1}$, and break up the matrices $\Phi^\pm$ into blocks as follows:

$$\Phi^\pm = \begin{pmatrix} \Phi_{11}^\pm & \Phi_{10}^\pm \\ \Phi_{01}^\pm & \Phi_{00}^\pm \end{pmatrix},$$

(the upper left blocks are $l^\mp \times l^\pm$).

THEOREM 2.7. *The operator* $W_{K,\Omega}$ *is d-normal if and only if*

(i) *(2.4) holds, and all the partial AP-indices of the matrix functions* $\tilde{K}_\pm$ *are nonnegative,*

(ii) *for all numbers* μ *lying on the ray*

$$\Sigma = \{\mu = \rho \exp(\pi i(\sigma + s - \frac{2}{p})) : \quad \rho \geq 0\}$$

the rows of the matrix

$$\left[\Phi_{00}^+ \Phi_{00}^- - \mu I, \ \Phi_{01}^+, \ \Phi_{00}^+ \Phi_{01}^- \right]$$

are linearly independent.

THEOREM 2.8. *The operator* $W_{K,\Omega}$ *is n-normal if and only if*

(i) *(2.4) holds, and all the partial AP-indices of the matrix functions* $\tilde{K}_\pm$ *are nonpositive,*

(ii) *for all* $\mu \in \Sigma$ *the columns of the matrix*

$$\begin{bmatrix} \Phi_{00}^+ \Phi_{00}^- - \mu I \\ \Phi_{10}^- \\ \Phi_{10}^+ \Phi_{00}^- \end{bmatrix}$$

are linearly independent.

If $0, m$ are both in J or in $\overline{J}$, then $N_0^{-1} N_m = I$, and therefore $\det \tilde{K}_\pm = $ constant. This means that $\lambda_1^\pm + \cdots + \lambda_{mn}^\pm = 0$, so that the partial indices of $\tilde{K}_\pm$ cannot be of the same sign without being equal 0. In particular, then the following statement follows from here and Theorems 2.6–2.8.

COROLLARY 2.9. *Suppose that* Ω *is bounded, the operator* $W_{K,\Omega}$ *is semi-Fredholm, and the corresponding local representatives* $\tilde{K}_\pm$ *are* AP_p-*factorable. Then* $W_{K,\Omega}$ *is Fredholm.*

The last theorem of this section provides an example of an explicit result obtained on a basis of Theorem 2.6 and Corollary 2.9, when an AP_p-factorization of $\tilde{K}_\pm$ can be constructed.

THEOREM 2.10. *Let Ω be a union of l bounded intervals, and in (2.1)*

(2.13) $$N \in C_p(\overline{\mathbb{R}}).$$

Then the following statements are equivalent:

   (i) $W_{K,\Omega}$ *is semi-Fredholm,*

  (ii) $W_{K,\Omega}$ *is Fredholm,*

 (iii) (a) $\operatorname{Re}\alpha = \sigma - s$,

         (b) $N_{\pm}\ (= N(\pm\infty))$ *are nonsingular, and*

         (c) *for all the eigenvalues $\eta_j\ (j = 1,\dots,n)$ of the matrix $N_-^{-1}N_+$*

$$\{\frac{1}{q} + \frac{\sigma + s}{2} + s - \nu_j\},\ \{\frac{1}{q} + \frac{\sigma + s}{2} + \sigma + \nu_j\} \neq 0,$$

where $\nu_j = \dfrac{1}{2\pi}\arg\eta_j,\ j = 1,\dots,n$. In addition,

(2.14) $$\operatorname{ind} W_{K,\Omega} = -l\sum_{j=1}^{n}\left([\frac{1}{q} + \frac{\sigma + s}{2} + s - \nu_j] + [\frac{1}{q} + \frac{\sigma + s}{2} + \sigma + \nu_j]\right).$$

Here, as usual, $[x]$ and $\{x\}$ stand for the integer part and fractional part of x, respectively.

According to Theorem 2.10, in case (2.13) the Fredholm criterion for the operator $W_{K,\Omega}$ does not depend on the number, sizes and disposition of the intervals forming Ω, but on the behavior of its presymbol at $\pm\infty$ only. Moreover, the index of the operator $W_{K,\Omega}$ equals its index on any finite interval multiplied by the number of intervals forming Ω. Of course, these properties depend heavily on (2.13).

The case of two intervals with rationally dependent endpoints for $n = 1$, $p = 2$ was considered in [3].

3. Auxiliary results.

3.1. Function classes. According to Stechkin's theorem ([34], see also [7, Theorem 2.11]), $\mathcal{M}_p$ contains all the functions of bounded variation, and, moreover,

(3.1) $$\|a\|_p \leq \|S\|_p(\|a\|_\infty + V_1(a)).$$

Here $\|a\|_\infty$ and $V_1(a)$ is a norm of a in $L_\infty(\mathbb{R})$ and its total variation correspondingly, $\|S\|_p$ is the norm in $L_p(\mathbb{R})$ of the operator (1.6), which coincides (see, for example, [18, Theorem 5.2]) with $\max\{\tan\dfrac{\pi}{2p}, \cot\dfrac{\pi}{2p}\}$.

Let us denote by $C_p(\mathbb{R})$ and $C_p(\dot{\mathbb{R}})$ the closure in $\mathcal{M}_p$ of the functions with finite total variation from $C(\mathbb{R})$ and $C(\dot{\mathbb{R}})$ correspondingly.

Some properties of the introduced classes are listed in the following

LEMMA 3.1.

(i) $C_p(\dot{\mathbb{R}}) = C_p(\overline{\mathbb{R}}) \cap C(\dot{\mathbb{R}})$;

(ii) $C_p(\overline{\mathbb{R}}) = \Pi C_p^\infty(\mathbb{R})$;

(iii) if $f \in AP_p, g \in C_p(\dot{\mathbb{R}})$ and $g(\infty) = 0$, then $fg \in C_p(\dot{\mathbb{R}})$;

(iv) for any $a \in SAP_p$:

(3.2)
$$a = a_+ u + a_-(1 - u) + a_0,$$

where $a_\pm \in AP_p, a_0 \in C_p(\overline{\mathbb{R}}), a_0(\pm\infty) = 0$, and u is a continuous on $\mathbb{R}$ function, increasing monotonically from 0 to 1. Moreover, the mappings $a \to a_\pm$ are homomorphisms of SAP_p onto AP_p ;

(v) if $\mathcal{B}$ is any of the algebras $AP, SAP, C(\overline{\mathbb{R}}), C(\dot{\mathbb{R}})$, and $\mathcal{B}_p$ is the corresponding algebra with the subscript p, then

(3.3)
$$\mathcal{B}_p = \mathcal{B}_q \subset \mathcal{B}_2 = \mathcal{B},$$

and

(3.4)
$$(\mathcal{B}_p)^{-1} = \mathcal{B}_p \cap (L_\infty)^{-1}$$

(symbol $\mathcal{Z}^{-1}$ stands, as usual, for a group of invertible elements of the algebra $\mathcal{Z}$).

Proof. Relations (3.3) follow from the property

(3.5)
$$\|\cdot\|_p = \|\cdot\|_q \geq \|\cdot\|_2 = \|\cdot\|_\infty$$

of $\mathcal{M}_p$-norms (see, for example, [7]). In particular, it follows from here that $C_p(\dot{\mathbb{R}}) \subset C(\dot{\mathbb{R}})$, which proves that the left side of *(i)* is contained in the right side. To prove the opposite inclusion, let us consider an arbitrary function $f \in C_p(\overline{\mathbb{R}})$ such that $f(+\infty) = f(-\infty)$. Let $\{f_n\} \subset C(\overline{\mathbb{R}})$ be a sequence of functions with the bounded variation which is convergent to f in the $\mathcal{M}_p$ norm. According to (3.5), f_n converges to f uniformly. Hence, in particular, $f_n(\pm\infty) \to f(\infty)$. Let us introduce a sequence of monotone on $\mathbb{R}$ functions u_n such that $u_n(+\infty) = f_n(+\infty) - f_n(-\infty), u_n(-\infty) = 0$. Due to (3.1), this means that $\|u_n\|_p \to 0$, anf therefore the sequence $\{f_n - u_n\}(\subset C_p(\dot{\mathbb{R}}))$ converges to f as well. Therefore, $f \in C_p(\dot{\mathbb{R}})$, which concludes the proof of *(i)*.

(ii) According to the theorem 1.16 in [12], $\mathcal{M}_r \cap C(\dot{\mathbb{R}}) \subset C_p(\dot{\mathbb{R}})$ for $r \in R_p$ given by (1.4). Therefore, $\mathcal{M}_r \cap C(\overline{\mathbb{R}}) \subset C_p(\overline{\mathbb{R}})$, and hence $C_p(\overline{\mathbb{R}})$ coincides with $\Pi C_p^\infty(\mathbb{R})$.

(iii) Suppose at first that f_0 is an almost periodic polynomial (and therefore its variation on any finite segment of $\mathbb{R}$ is bounded), and g_0 is a continuous piecewise linear function with a compact support. Then $f_0 g_0$ is a function with a compact support which is continuous on $\mathbb{R}$ and vanishing at ∞. Hence, $f_0 g_0 \in C_p(\dot{\mathbb{R}})$. It remains to notice that an arbitrary function $g \in C_p(\dot{\mathbb{R}})$ vanishing at the infinity can be represented as a limit in $\mathcal{M}_p$

of functions of the type g_0 ([7, remark 2.12]), and any $f \in AP_p$ is a limit of almost periodic polynomials according to the definition of AP_p.

(iv) Let us consider the set of all functions a having representation (3.2). Observe first of all that if along with u, u_1 is another continuous function which is monotonically increasing from $u_1(-\infty) = 0$ to $u_1(+\infty) = 1$, then $a = a_+ u_1 + a_-(1 - u_1) + a'_0$, where $a'_0 = (a_+ - a_-)(u - u_1) + a_0 \in C_p(\dot{\mathbb{R}})$ according to (ii), and $a'_0(\infty) = 0$. Therefore, without loss of generality the function u may be supposed to be the same for all representations (3.2). The uniqueness of $a_\pm$ for $a \in SAP_p$ follows from the inclusion $SAP_p \subset SAP$.

Let $b = b_+ u + b_-(1 - u) + b_0$ be, along with a, another function in SAP_p. Then

$$a + b = (a_+ + b_+)u + (a_- + b_-)(1 - u) + (a_0 + b_0), \quad ab = a_+ b_+ u + a_- b_-(1 - u) + c_0,$$

where

$$\begin{aligned}
c_0 - a_0 b_0 &= (a_+ b_+ + a_- b_-)(u - u^2) \\
&\quad + (a_+ b_- + a_- b_+)u(1 - u) + a_0(b_+ u + b_-(1 - u)) + b_0(a_+ u + a_-(1 - u)).
\end{aligned}$$

All the summands in the expression for $c_0 - a_0 b_0$ are in $C_p(\dot{\mathbb{R}})$ according to (ii), and vanish at ∞. As a result, the set of functions (3.2) form an algebra $\mathcal{A}$, and mappings $a \mapsto a_\pm$ are the algebraic homomorphisms of $\mathcal{A}$ onto AP_p.

To complete the proof of (iv) it remains to show that

$$(3.6) \qquad\qquad (\forall a \in \mathcal{A}) \quad \|a_\pm\|_p \leq \|S\|_p \|a\|_p,$$

since it will follow from here that $\mathcal{A}$ is closed in $\mathcal{M}_p$, and therefore coincides with the whole SAP_p.

Let us prove (3.6) for a_+, the proof for a_- being identical. Since the set of almost periods of $a_+ \in AP_p (\subset AP)$ is relatively dense ([20, p. 20]), there exists a sequence $h_n \in \mathbb{R}$ such that $h_n \uparrow +\infty$ and $\|a_+ - a_+^{(h_n)}\|_\infty < n^{-1}$ (we use an abbreviation $f^{(h)}(x) = f(x + h)$). Then for $p \in [2, \infty]$

$$(3.7) \qquad e^{-ixh_n} W_{\chi_c a_+} e^{ixh_n} I = W_{\chi_{c-h_n} a_+^{(h_n)}} \xrightarrow{s} W_{a_+} \quad (n \to \infty),$$

where χ_c is a characteristic function of the interval (c, ∞). Indeed, according to Hausdorff-Young theorem (see [7, Theorem 1.32]) for rapidly decreasing functions $f \in S(\mathbb{R})$

$$\|W_{\chi_{c-h_n} a_+^{(h_n)}} - W_{a_+} f\|_{L_p(\mathbb{R})} \leq$$

$$\|\mathcal{F}^{-1}\|_{L_q(\mathbb{R}) \to L_p(\mathbb{R})} \left(\|a_+^{(h_n)} - a_+\|_\infty \|\chi_{c-h_n}\|_\infty \|\mathcal{F}f\|_{L_q(\mathbb{R})} + \|a_+\|_\infty \|(\chi_{c-h_n} - 1)\mathcal{F}f\|_{L_q(\mathbb{R})} \right),$$

and therefore $\lim\limits_{n\to\infty} W_{\chi_{c-hn}a_+^{(hn)}} f = W_{a_+} f$. The latter property together with density of $\mathcal{S}(\mathbb{R})$ in $L_p(\mathbb{R})$ implies (3.7). From (3.7) it follows that

$$\|a_+\|_p = \|W_{a_+}\| \le \liminf_{n\to\infty} \|e^{-ixh_n} W_{\chi_c a_+} e^{ixh_n} I\| = \|W_{\chi_c a_+}\| \le$$
$$\|W_{\chi_c a}\| + \|W_{\chi_c(a_- - a_+)(1-u)}\| + \|W_{\chi_c a_0}\| \le$$
$$\|W_{\chi_c}\| \|a\|_p + \|a_- - a_+\|_p \|\chi_c(1-u)\|_p + \|\chi_c a_0\|_p.$$

Since according to (3.1),

$$\|\chi_c(1-u)\|_p \le \|S\|_p \left(|1 - u(c)| + 2|1 - u(c)|\right),$$

and a_0 can be approximated in $\mathcal{M}_p$ by continuous piecewise linear functions with compact support, for all $\varepsilon > 0$ there exists $c > 0$ such that

$$|1 - u(c)| < \varepsilon/(6\|S\|_p \|a_- - a_+\|_p + 1), \quad \|\chi_c a_0\|_p < \varepsilon/2.$$

Then $\|a_- - a_+\|_p \|\chi_c(1-u)\|_p + \|\chi_c a_0\|_p < \varepsilon$. Finally, $\|a_+\|_p \le \|W_{\chi_c}\| \|a\|_p + \varepsilon$, and, due to the arbitrariness of ε, $\|a_+\|_p \le \|W_{\chi_c}\| \|a\|_p$. It remains only to take into consideration the fact that

$$\|W_{\chi_c}\| = \|P_+\| \le (1 + \|S\|_p)/2 \le \|S\|_p.$$

If $1 < p < 2$, then, taking adjoints and using (3.5):

$$\|a_+\|_p = \|a_+\|_q \le \|S\|_q \|a\|_q = \|S\|_p \|a\|_p.$$

This completes the proof of (3.6), and, at the same time, of the statement *(iv)*.

Let us turn now to property (3.4). For $\mathcal{B} = AP$ it is proved in [10, p. 239], and for $\mathcal{B} = C(\overline{\mathbb{R}})$ it follows from [30, Theorem 3.9], according to which a compact of maximal ideals of the algebra $C_p(\overline{\mathbb{R}})$ coincides with $\overline{\mathbb{R}}$. According to *(i)*, (3.4) holds also for $\mathcal{B} = C(\dot{\mathbb{R}})$. It remains only to consider the case $\mathcal{B} = SAP$.

So, let a be a function of the form given in (3.2) such that $a^{-1} \in L_\infty$. Then the corresponding $a_\pm$ are bounded away from 0 in the neighborhood of $\pm\infty$. Being almost periodic, they are separated from 0 on $\mathbb{R}$. In other words, $a_\pm^{-1} \in L_\infty$, and so by (3.4) for the case $\mathcal{B} = AP$, $a_\pm^{-1} \in AP_p$. Put $b = a_+^{-1}u + a_-^{-1}(1 - u)$. Then $b \in SAP_p$ and is bounded away from 0 outside of a certain interval $J = [-\zeta, \zeta]$ together with $a_\pm^{-1}$. Let b_0 be a function continuous on $\mathbb{R}$ with bounded variation having its support in J and such that $b + b_0$ does not vanish on J. Then $b + b_0 \in SAP_p$ and $(b + b_0)^{-1} \in L_\infty$. According to *(iii)*, the almost periodic components of the function $c = a(b + b_0) \in SAP_p$ at $\pm\infty$ are $c_\pm = 1$, and therefore $c \in C_p(\dot{\mathbb{R}})$. Further, $c^{-1} = a^{-1}(b + b_0)^{-1} \in L_\infty$. By (3.4) with $\mathcal{B} = C_p(\dot{\mathbb{R}})$, it follows that $c^{-1} \in C_p(\dot{\mathbb{R}})$. But then $a^{-1} = c^{-1}(b + b_0) \in SAP_p$. $\blacksquare$

REMARK. It can be shown that the constant $||S||_p$ in (3.6) can be replaced by 1.

For $p = 2$ the crucial role in the Fredholm theory of operators W_K ($K \in SAP$) is played by the Douglas algebras $H_\infty^\pm + C(\dot{\mathbb{R}})$, where $H_\infty^\pm$ is a Hardy class of functions analytic and bounded in the upper/lower half-plane. In particular, the following result of Sarason [28] is extensively used.

LEMMA 3.2. If $u \in C(\overline{\mathbb{R}})$, $h \in H_\infty^\pm$, and $\lim_{y \to \pm\infty} h(iy) = 0$, then $uh \in H_\infty^\pm + C(\dot{\mathbb{R}})$.

In our setting the Douglas algebras $H_\infty^\pm$ must be replaced by $\mathcal{A}^\pm$, the closed in $\mathcal{M}_p$ algebras generated by $H_\infty^\pm \cap \mathcal{M}_p$ and $C_p(\dot{\mathbb{R}})$. The Sarason Lemma is replaced then by

LEMMA 3.3. If $h \in AP_p^\pm$ and $\mathbf{M}(h) = 0$, then $uh \in \mathcal{A}^\pm$ for all $u \in C_p(\overline{\mathbb{R}})$.

Proof. Let us consider the case $h \in AP_p^+$. It suffices to prove the desired statement for $h(x) = e^{i\nu x}$ ($\nu > 0$), since any function $h \in AP_p^+$ with $\mathbf{M}(h) = 0$ is a limit in $\mathcal{M}_p$ of almost periodic polynomials with positive Fourier coefficients, and $\mathcal{A}^+$ is a closed subalgebra of $\mathcal{M}_p$. For such a choice of h let us represent a given function $u \in C_p(\overline{\mathbb{R}})$ in a form

$$(3.8) \qquad u(x) = \frac{u(-\infty) + u(+\infty)}{2} + \frac{u(+\infty) - u(-\infty)}{2} \mathrm{Si}_3(\nu x) + v(x),$$

where

$$\mathrm{Si}_3(\zeta) = -\pi \int_0^\zeta (\xi^2 - \pi^2)^{-1} \xi^{-1} \sin \xi \, d\xi$$

is a modified integral sine.

Obviously, the derivative of Si_3 is absolutely integrable, hence Si_3 itself is a continuous on $\overline{\mathbb{R}}$ function with bounded variation and, in particular, $\mathrm{Si}_3 \in C_p(\overline{\mathbb{R}})$. It is important for what follows that $\mathrm{Si}_3(\pm\infty) = \pm 1$ and $e^{ix}\mathrm{Si}_3(x) \in H_\infty^+$ (see, for example,[13].)

The function v defined by (3.8) belongs to $C_p(\overline{\mathbb{R}})$ together with u and Si, and $v(\pm\infty) = 0$. Hence, the statements (i) and (ii) of Lemma 3.1 imply that $hv \in C_p(\dot{\mathbb{R}})$. The other summands in the right side of (3.8) are sent to $H_\infty^+ \cap \mathcal{M}_p$ under multiplication by h. Therefore, $hu \in \mathcal{A}^+$. ∎

LEMMA 3.4. Suppose that F, G are invertible matrix functions from SAP_p, and their local representatives at $\pm\infty$ are connected by the relations $F_\pm = N_\pm^+ G_\pm N_\pm^-$, where $(N_+^+)^{\pm 1}, (N_-^+)^{\pm 1} \in AP_p^+$, $(N_+^-)^{\pm 1}, (N_-^-)^{\pm 1} \in AP_p^-$, and

$$\mathbf{M}(N_+^+) = \mathbf{M}(N_-^+), \ \mathbf{M}(N_+^-) = \mathbf{M}(N_-^-).$$

Then $F = N^+ G N^-$, where the matrix function $N^\pm$ is homotopic to an invertible element of $C_p(\dot{\mathbb{R}})$ in the set $(\mathcal{A}^\pm \cap SAP_p)^{-1}$.

The proof of Lemma 3.4 goes exactly the same way as the proof of Lemma 3.2 in [16]. The only difference is that the approximation of $N_\pm^\pm$ by almost periodic polynomials must be carried out now in the topology of $\mathcal{M}_p$, and Lemma 3.3 is used instead of Sarason Lemma 3.2.

3.2. Operators on L_p. For any two square matrices $A, B \in SAP_p$ of the same size, let us consider the so called *pair operator*

$$(3.9) \qquad R_{A,B} = W_A^0 \chi_+ + W_B^0 \chi_- \, (= \mathcal{F}^{-1} A \mathcal{F} \chi_+ + \mathcal{F}^{-1} B \mathcal{F} \chi_-) : L_p(\mathbb{R}) \to L_p(\mathbb{R}),$$

where $\chi_\pm$ is a characteristic function of $\mathbb{R}_\pm$.

Of course, $R_{K,I}$ has a triangular block representation

$$R_{K,I} = \begin{bmatrix} W_K & 0 \\ * & I \end{bmatrix}$$

with respect to the direct decomposition $L_p(\mathbb{R}) = L_p(\mathbb{R}_+) \dotplus L_p(\mathbb{R}_-)$, and therefore the Fredholm properties of $R_{K,I}$ and W_K are the same. Nevertheless, for technical reasons we shall need to consider the more general case of $R_{A,B}$ with $B \neq I$.

LEMMA 3.5. *In order for the operator $R_{A,B}$ with $A, B \in SAP_p$ to be (semi-) Fredholm it is necessary that*

$$(3.10) \qquad\qquad\qquad A^{-1}, B^{-1} \in SAP_p.$$

Proof. Let us concentrate on the case of n-normal $R_{A,B}$ (d-normality can be considered then by passing to adjoint operator $R_{A,B}^*$ acting in $L_q(\mathbb{R})$.) Suppose that $A^{-1} \notin SAP_p$. According to statement *(v)* of Lemma 3.1 it means that $\inf_{x \in \mathbb{R}} |\det A(x)| = 0$. Let us show that in this case in any neighborhood of $R_{A,B}$ there can be found an operator $R_{M,B}$ where the matrix function $M (\in SAP_p)$ is constant and singular on an interval of $\mathbb{R}$. To this end, for any fixed $\delta_1 > 0$, pick a matrix function A_1 such that $\|R_{A_1,B} - R_{A,B}\| < \delta_1$ and $A_1 \in SAP_r$ for all $r \in (1, \infty)$ (this is possible according to the definition of SAP_p). Since $\|A - A_1\| < \delta_1$, for all the eigenvalues $\lambda_j(x)$ of $A_1(x)$,

$$(3.11) \qquad\qquad\qquad \inf_{j,x} |\lambda_j(x)| < \delta_2,$$

where δ_2 can be made arbitrarily small by appropriate choice of δ_1. According to (3.11), in the δ_2-neighborhood of A_1 there exists a matrix function A_2 of the form $A_1 - cI$ vanishing somewhere on $\mathbb{R}$. Suppose, $M_0 = K_2(x_0)$ is singular. Introduce a smooth function φ_ε which is equal 1 in the $\frac{\varepsilon}{2}$-neighborhood of x_0, 0 exterior to its ε-neighborhood, and monotone at each side of x_0. Put $M = A_2 + \varphi_\varepsilon(M_0 - A_2)$. By (3.1), $\varphi_\varepsilon \in \mathcal{M}_r$ for all $r \in (1, \infty)$, and the constant C_r in the inequalities $\|\varphi_\varepsilon\|_r \leq C_r$ does not depend on ε. The matrix function M belongs to $\mathcal{M}_r$ together with A_2 and φ_ε. Of course, M coincides with M_0 in the $\frac{\varepsilon}{2}$-neighborhood of x_0, and therefore is constant and singular on this interval. It remains to show that ε can be chosen so that the norm $\|R_{M,B} - R_{A_2,B}\| \left(= \|W_{\varphi_\varepsilon(M_0 - A_2)}\| \right)$ is arbitrarily small. Clearly it suffices to prove the corresponding property for all the scalar operators $W_{\varphi_\varepsilon z_{ij}}$, where z_{ij} is (i,j)-th element of the matrix $M_0 - K_2$. Choose $r \in (1, \min\{p,q\})$. It follows from [7, Proposition 2.5] (see also [12]) that for $\gamma = r|p-2|/p|r-2|$:

$$||\varphi_\varepsilon z_{ij}||_p \leq ||\varphi_\varepsilon z_{ij}||_\infty^{1-\gamma} ||\varphi_\varepsilon z_{ij}||_r^\gamma \leq C_r^\gamma ||z_{ij}||_r^\gamma \text{ ess sup}\{|z_{ij}(x)|^{1-\gamma} : |x - x_0| < \varepsilon\},$$

and the desired estimate follows.

Hence, for all $\delta > 0$, by choosing appropriate values of δ_1 and then ε, it is always possible to fulfil the requirement $||R_{M,B} - R_{K,B}|| < \delta$. If the given operator $R_{K,B}$ is n-normal, then according to the stability property it may be assumed that W_M is n-normal as well.

Since M_0 is singular, there exists a matrix $D \neq 0$ such that $M_0 D = 0$. Put $X = \varphi_{\varepsilon/2} D$. Then

(3.12) $$R_{M,B} X_+ W_X \chi_+ = W_{MX}^0 \chi_+ - W_M^0 X_- W_X^0 \chi_+.$$

The operator $\chi_- W_X^0 \chi_+ = D\chi_- W_{\varphi_{\varepsilon/2}}^0 \chi_+$ is compact, and $W_{MX}^0 = \mathcal{F}^{-1} M D\varphi_{\varepsilon/2}\mathcal{F} = 0$ since in the $\frac{\varepsilon}{2}$-neighborhood of x_0, $MD = M_0 D = 0$. Therefore, the left side of (3.12) is compact as well. Due to [17, p. 26], from the n-normality of the left factor $RW_{M,B}$ and compactness of the product it follows that the right factor $W_X = DW_{\varphi_{\varepsilon/2}}$ is also compact. The contradiction obtained completes the proof of the lemma. ■

According to Lemma 3.5, discussing (semi-)Fredholmness of operators $R_{A,B}$ from now on without loss of generality we may suppose that (3.10) is satisfied. But then $R_{A,B}$ differs from $R_{B^{-1}A,I}$ by an invertible multiple of W_B^0, and therefore the Fredholm properties of $R_{A,B}$ coincide with those of W_K, where $K = B^{-1}A$ is an invertible element of SAP_p.

In the rest of this section we will deal with operators W_K only.

LEMMA 3.6. *If $a,b \in M_p$ and $a \in \mathcal{A}^+$ or $b \in \mathcal{A}^-$, then the operator $Z = W_a W_b - W_{ab}$ is compact in $L_p(\mathbb{R}_+)$.*

Proof. For $a \in H_\infty^+ \cap M_p$ the exact equality $W_a W_b = W_{ab}$ holds (see [7, Proposition 2.7] [1]. If $a \in C_p(\dot{\mathbb{R}})$, then the operator $T = \chi_+ \mathcal{F}^{-1} a\mathcal{F} - \mathcal{F}^{-1} a\mathcal{F}\chi_+$ is compact due to [7, Lemma 7.3], and $W_a W_b - W_{ab} = \chi_+ T\mathcal{F}^{-1} b\mathcal{F}$. Since

$$W_{a_1+a_2} W_b - W_{(a_1+a_2)b} = W_{a_1} W_b - W_{a_1 b} + W_{a_2} W_b - W_{a_2 b},$$

$$W_{a_1 a_2} W_b - W_{a_1 a_2 b} = (W_{a_1 a_2} - W_{a_1} W_{a_2})W_b + W_{a_1}(W_{a_2} W_b - W_{a_2 b})$$
$$+ (W_{a_1} W_{a_2 b} - W_{a_1 a_2 b}),$$

sums of products of elements from $H_\infty^+ \cap M_p$ and $C_p(\dot{\mathbb{R}})$ also have the desired property. For an arbitrary $a \in \mathcal{A}^+$ the compactness of Z follows from here by a passage to the limit. The case $b \in \mathcal{A}^-$ can be considered analogously. ■

---

[1] Here and in what follows difference in the order of multiplication as compared with [7] is caused by a notational difference: our operator $\mathcal{F}$ is what is called $\mathcal{F}^{-1}$ in [7].

COROLLARY 3.7. *If* $a^{\pm 1} \in \mathcal{A}^+, b^{\pm 1} \in \mathcal{A}^-$, *then operators* W_a, W_b *are Fredholm in* $L_p(\mathbb{R}_+)$.

Indeed, $W_{a^{-1}}$ and $W_{b^{-1}}$ are two-sided regularizers for W_a and W_b correspondingly.

LEMMA 3.8. *Under the conditions of Lemma 3.4 the operator* W_G *is d-normal (n-normal, Fredholm) if and only if* W_K *is, and*

(3.13) $$\operatorname{ind} W_K = \operatorname{ind} W_{N+} + \operatorname{ind} W_{N-} + \operatorname{ind} W_G.$$

Proof. The same reasoning can be applied as in the proof of Corollary 3.1 in [16]. Of course, Corollary 3.7 must be used instead of the corresponding results on the Fredholmness of singular integral operators T_G with symbols G invertible in $H_\infty^\pm + C$. ∎

In the case of AP_p-factorable local representatives $K_\pm$ of a presymbol $K \in SAP_p$ the Lemma 3.8 allowes one to reduce the question of (semi-)Fredholmness of W_F to the corresponding question for convolution operators with scalar presymbols ($\in SAP_p$) and matrix presymbols from $C_p(\overline{\mathbb{R}})$ (see Lemma 3.3 and Theorem 3.1 in [16], where such a reduction has been carried out for operators T_F). Since the (semi-)Fredholm theory for operators W_f, W_G with scalar $f \in SAP_p$ and $G \in C_p(\overline{\mathbb{R}})$ is known (see [8] and [6]), this means that the results on (semi-)Fredholmness of operators W_F may now be obtained in the same way as in Theorems 3.2–3.4 of [16]. The corresponding results match the ones claimed by Theorems 2.6, 2.7, 2.8, if one takes into consideration the fact that in our current setting $m = 1$, $\alpha = \sigma = s = 0$, $N_0 = I$, and therefore $\tilde{K}$ coincides with K.

Of course, Theorems 2.6, 2.7 and 2.8 in their full generality automatically follow from this particular case as soon as Theorem 2.1 is established. This will be done in section 4.1.

3.3. AP-factorization. Factorization of almost periodic matrix functions G is quite a complicated subject, and explicit conditions of existence, formulas for partial indices and $d(G)$ are difficult to obtain. Some results in this direction can be found in [16, Section 2]. See [32] for further developments, and [15] for (still) unsolved problems.

Here we will restrict ourselves to the relatively simple case when

(3.14)
$$G(x) = \begin{bmatrix} c_1 e^{i\lambda_1 x} & 0 & \cdots\cdots\cdots & 0 \\ 0 & c_2 e^{i\lambda_2 x} & & 0 \\ \vdots & & \ddots & \vdots \\ 0 & \cdots\cdots\cdots & c_{m-1} e^{i\lambda_{m-1} x} & 0 \\ d_1 e^{i\mu_1 x} & d_2 e^{i\mu_2 x} & \cdots & d_{m-1} e^{i\mu_{m-1} x} \quad c_m e^{i\mu_0 x} \end{bmatrix},$$

where $c_1, \ldots, c_m, d_1, \ldots, d_{m-1} \in \mathbb{C}^{n \times n}$. The condition

(3.15) $$\det(c_1 \cdots c_m) \neq 0$$

is equivalent to the invertibility of G, and is therefore necessary for its AP-factorability. We suppose in what follows that this condition is met. In [33, Theorem 2] there is proposed a recursive (on m) construction of AP_W-factorization of (3.14) in the case when the blocks d_j have the "extreme rank" property, that is, each of them either is invertible or equals zero. This property is obviously satisfied for matrix functions (3.14) with $n = 1$, and hence these matrices are AP_W-factorable whenever $c_1 \cdots c_m \neq 0$.

The general case of matrices (3.14) with $n > 1$ requires a scrupulous analysis, which we are not going to carry out here. For our purposes it suffices to consider the case when

(3.16)   the set $\{d_1, \ldots, d_{m-1}, c_m\}$ contains not more than two different elements.

We will denote by d the element of $d_1, \ldots, d_{m-1}$ different from $c = c_m$ (if any), and put $d = c$ if $d_1 = \ldots = d_{m-1} = c$.

Of course, (3.16) is no restriction when $m \leq 2$ — the case which was dealt with in [16, Section 2.3].

LEMMA 3.9. *(i) Suppose that for the matrix function G given by (3.14), conditions (3.15) and (3.16) are satisfied. Then G is AP_W-factorable.*

(ii) If, in addition,

(3.17)
$$\lambda_j > 0 \ (j = 1, \ldots, m-1) \ and$$
$$\mu_j = -(\lambda_{j+1} + \cdots + \lambda_{m-1}) \ (j = 0, \ldots, m-2), \ \mu_{m-1} = 0,$$

then for partial AP-indices of G to be of the same sign it is necessary and sufficient that

(3.18) $\det d \neq 0.$

In the latter case all the partial AP-indices of G equal zero, and

(3.19) $$d(G) = \begin{bmatrix} 0 & \cdots & 0 & (-1)^{m-1} c_1 d_1^{-1} c_m \\ (-1)^m c_2 d_2^{-1} d_1 & 0 & \ddots & 0 \\ 0 & \ddots & \ddots & \vdots \\ 0 & \cdots & d_{m-1} & 0 \end{bmatrix},$$

where the $(j, j-1)$-block equals $(-1)^{m+j} c_j d_j^{-1} d_{j-1}$ for $j = 2, \ldots, m-1$, and all other blocks not shown in (3.19) equal 0.

Proof. *(i)* Multiplying (3.14) by constant nonsingular matrices $\operatorname{diag}[h, \ldots, h]$ and $\operatorname{diag}[c_1 h, \ldots, c_m h]^{-1}$ from the right and left correspondingly, we can suppose without loss of generality that

(3.20) $c_1 = \ldots = c_m = I,$

and the matrix d is reduced to its Jordan canonical form $h^{-1}dh$. Under these circumstances matrix G is permutation similar to the direct sum of matrices of the form (3.14), with d in them being basic Jordan blocks $J_{\zeta,k}$.

Therefore, it suffices to prove AP_W-factorability of matrices (3.14) satisfying (3.20) and $d_j = J_{\zeta,k}$ or 0. For $\zeta \neq 0$ all such d_j are invertible, and for $\zeta = 0$ a further permutation of G reduces it to a matrix G_0 of the same type (3.14) with $n = 1$. In both cases AP_W-factorability follows from the results of [33] mentioned above. Let us mention that the block G_0 contains a direct summand $e^{i\lambda_j x}$ for each $d_j = J_{0,k}$, and therefore the corresponding λ_j are among the partial AP-indices of G.

(ii) According to (3.17), $\det G(x) = \text{const}$, and therefore the sum of its partial indices equals 0. So, they are of the same sign if and only if all of them equal zero. If $\det d = 0$, then the block G_0 is actually present in the direct decomposition of G discussed in the first step of the proof, and so some of the partial AP-indices of G equal λ_j ($\neq 0$). This proves the necessity of condition (3.18).

If on the other hand (3.18) is satisfied, then all the blocks d_j in (3.14) are nonsingular. In this case the AP_W-factorization of (3.14) satisfying (3.15), (3.17) was obtained in [31]. According to Lemma 1 of [31], all the partial AP-indices equal 0, and $d(G)$ is given by (3.19). ∎

4. Operators $W_{K,\Omega}$.

4.1. Proof of Theorem 2.1. Let $\{\eta_j\}_{j=0}^m$ be a partitioning of unity such that $\eta_j \in C^\infty(\mathbb{R})$, $\eta_j(x) = 1$ in the neighborhood of Ω_j, and $\eta_j(x) = 0$ outside the union of Ω_j with intervals adjacent to it.

It is always possible to choose these η_j to be in the form $\eta_{j1} - \eta_{j2}$, where η_{jk} are monotonic C^∞-functions with compact support. As in the proof of Theorem 7.3 in [21] it can be established that multiplication by such functions η_{jk} (and therefore η_j) is a bounded operator in all the spaces $\mathcal{H}_{s,p}(\mathbb{R})$ ($s \in \mathbb{R}, 1 < p < \infty$).

Hence, the following direct decompositions hold:

(4.1)
$$\overset{0}{\mathcal{H}}_{\sigma,p}(\Omega) = \overset{\circ}{\sum_{j \in J}} \overset{0}{\mathcal{H}}_{\sigma,p}(\Omega_j),$$

(4.2)
$$\overset{0}{\mathcal{H}}_{s,p}(\mathbb{R} \setminus \Omega) = \overset{\circ}{\sum_{j \in \bar{J}}} \overset{0}{\mathcal{H}}_{s,p}(\Omega_j).$$

According to these decompositions,

$$\varphi = \sum_{j \in J} v_j, \quad u = \sum_{j \in \bar{J}} v_j,$$

where

(4.3)
$$v_j \in \overset{0}{\mathcal{H}}_{s_j,p}(\Omega_j),$$

s_j is given by (2.3), and the equation (1.2) can be rewritten as

(4.4)
$$\sum_{j=0}^{m} W_{K_j}^0 v_j = \tilde{f}$$

with

(4.5)
$$K_j = \begin{cases} K(=\mathcal{F}k), & j \in J \\ I_n, & j \in \bar{J} \end{cases}$$

and $\tilde{f}(\in \mathcal{H}_{s,p}(\mathbb{R}))$ being an arbitrary continuation of $f \in \mathcal{H}_{s,p}(\Omega)$. Let us consider equation (4.4) in more general setting, where instead of (2.1) and (4.5) we just demand that for every k_j there exists $\alpha_j (\in \mathbb{C})$ such that

(4.6)
$$N_j(x) \overset{\text{def}}{=} (x^2 + 1)^{-\alpha_j/2} \in SAP_p$$

(of course, (4.6) follows from (4.5) when α_j and s_j are given by (2.3).)

The operator

(4.7)
$$R = [W_{K_0}^0, \dots, W_{K_m}^0]: \sum_{j=0}^{m} \overset{0}{\mathcal{H}}_{s_j,p}(\Omega_j) \to \mathcal{H}_{s,p}(\mathbb{R})$$

corresponding to the equation (4.4) is bounded if in (4.6)

(4.8)
$$s \le s_j - \operatorname{Re}\alpha_j \quad (j = 0, \dots, m).$$

We suppose in what follows that condition (4.8) is satisfied.

LEMMA 4.1. *The operator R defined by (4.7) is Fredholm (n-normal, d-normal) exactly when the operator $R_{A,B}$ defined by (3.9) is, where the matrix coefficients A and B are given by*

(4.9)
$$A = \begin{bmatrix} I_n & \cdots & 0 & 0 \\ \vdots & \ddots & \vdots & \vdots \\ 0 & \cdots & I_n & 0 \\ a_1 & \cdots & a_{m-1} & a_m \end{bmatrix},$$

$$a_j(x) = \left(\frac{x-i}{x+i}\right)^{s-s_j+\overline{\alpha_j}/2} (x+i)^{s-s_j+\operatorname{Re}\alpha_j} e^{-ic_j-1^x} N_j(x) \quad (j = 1, \dots, m),$$

(4.10) $B(x) =$

$$\mathrm{diag}\left[\left(\frac{x-i}{x+i}\right)^{s_j+i\,\mathrm{Im}\,\alpha_j} e^{-i\lambda_j x} I_n\right]_{j=1}^{m-1} \oplus \left(\frac{x-i}{x+i}\right)^{s+\frac{\alpha_0}{2}} (x+i)^{s-s_0+\mathrm{Re}\,\alpha_0} e^{-ic_0 x} N_0(x).$$

Proof. Denote by V_λ the shift operator

$$(V_\lambda f)(x) = f(x+\lambda),$$

and put

$$v_j^+ = V_{c_{j-1}} v_j \qquad (j=1,\dots,m)$$

$$v_j^- = V_{c_j} v_j \qquad (j=0,\dots,m-1)$$

Then, of course,

(4.11) $$v_j^+ = V_{-\lambda_j} v_j^- \quad (j=1,\dots m-1),$$

and the condition (4.3) can be rewritten as

(4.12)
$$v_0^- \in \overset{0}{\mathcal{H}}_{s_0,p}(\mathbb{R}_-), \quad v_m^+ \in \overset{0}{\mathcal{H}}_{s_m,p}(\mathbb{R}_+),$$
$$v_j^\pm \in \overset{0}{\mathcal{H}}_{s_j,p}(\mathbb{R}_\pm), \quad (j=1,\dots,m)$$

The equation (4.4) in our new notations takes the form

(4.13)
$$\sum_{j=1}^m W_{K_j}^0 V_{-c_{j-1}} v_j^+ + W_{K_0}^0 V_{-c_0} v_0^- = \tilde{f}$$

For any $t,\omega \in \mathbb{R}$ let us put

$$X_{t,\omega}^\pm = \mathcal{F}^{-1}(x \mp i)^{t+i\omega} \mathcal{F}.$$

According to [7, Lemma 5.1], the operators $X_{t,0}^\pm$ map isomorphically $\overset{0}{\mathcal{H}}_{t,p}(\mathbb{R}_\pm)$ onto $L_p(\mathbb{R}_\pm)$ (note that these operators are isomorphisms of $\overset{0}{\mathcal{H}}_{t,p}(\mathbb{R})$ onto $L_p(\mathbb{R})$ simply due to the definition (1.1)). At the same time, with the help of Mikhlin's theorem on multiplicators (see, for example, [23, p. 239] it can be established that $X_{0,\omega}^\pm$ are endomorphisms of $L_p(\mathbb{R})$ and $L_p(\mathbb{R}_\pm)$. Since $X_{t,\omega}^\pm = X_{t,0}^\pm X_{0,\omega}^\pm$, the above mentioned isomorphic properties of $X_{t,0}^\pm$ are inherited by $X_{t,\omega}^\pm$.

Introduce the (vector) functions

$$\psi_j^\pm = \pm X_{s_j,\omega_j}^\pm v_j^\pm \quad (j=1,\dots,m-1)$$

$$\psi_m^+ = X_{s_m,\omega_m}^+ v_m^+, \quad \psi_m^- = X_{s_0,\omega_0}^- v_0^-,$$

and put

$$g = X_{s,0}^+ \tilde{f} \, (\in L_p(\mathbb{R})).$$

Then (4.12) is equivalent to

(4.14) $\psi_j^\pm \in L_p(\mathbb{R}_\pm) \quad (j = 1, \dots, m),$

and equations (4.11), (4.13) may be rewritten as

(4.15) $\psi_j^+ + \mathcal{F}^{-1} \left(\dfrac{x-i}{x+i} \right)^{s_j + i\omega_j} e^{-i\lambda_j x} \mathcal{F} \psi_j^- = 0, \quad (j = 1, \dots, m-1)$

and

(4.16)
$$\sum_{j=1}^m \mathcal{F}^{-1}(x-i)^{s-s_j-i\omega_j} e^{-ic_j-1^x} K_j(x) \mathcal{F}\psi_j^+ +$$
$$\mathcal{F}^{-1}(x-i)^s(x+i)^{-s_0-i\omega_0} e^{-ic_0 x} K_0(x)\mathcal{F}\psi_m^- \quad = \quad g$$

respectively.

Putting $\omega_j = \operatorname{Im}\alpha_j$ $(j = 0, \dots, m)$ and taking (4.6) into consideration, we can rewrite (4.16) as

$$\sum_{j=1}^m \mathcal{F}^{-1} \left(\frac{x-i}{x+i} \right)^{s-s_j+\frac{\overline{\alpha_j}}{2}} (x+i)^{s-s_j+\operatorname{Re}\alpha_j} e^{-ic_j-1^x} N_j(x)\mathcal{F}\psi_j^+ +$$
$$\mathcal{F}^{-1} \left(\frac{x-i}{x+i} \right)^{s+\frac{\alpha_0}{2}} (x+i)^{s-s_0+\operatorname{Re}\alpha_0} e^{-ic_0 x} N_0(x)\mathcal{F}\psi_m^- \quad = \quad g.$$

Finally, if we denote $\psi_j = \psi_j^+ + \psi_j^-$ $(\in L_p(\mathbb{R}))$, then the system of equations (4.15), (4.16) has the form

(4.17)
$$R_{A,B} \begin{bmatrix} \psi_1 \\ \vdots \\ \vdots \\ \psi_m \end{bmatrix} = \begin{bmatrix} 0 \\ \vdots \\ 0 \\ g \end{bmatrix}.$$

Therefore, equations (4.4) and (4.17) are equivalent.

Let us consider a direct decomposition $\mathcal{L}_1 + \mathcal{L}_2$ of the mn-size columns space $L_p(\mathbb{R})$, where $\mathcal{L}_1$ ($\mathcal{L}_2$) stands for the set of columns with the last n (correspondingly, first $n(m-1)$) elements equal 0 identically. The equivalence of (4.4) and (4.17) means that $\dim \operatorname{Ker} R = \dim \operatorname{Ker} R_{A,B}$, $\operatorname{Im} R$ is closed exactly when $\operatorname{Im} R_{A,B} \cap \mathcal{L}_2$ is closed, and $\operatorname{codim} \operatorname{Im} R = \dim \mathcal{L}_2/(\operatorname{Im} R_{A,B} \cap \mathcal{L}_2)$. Therefore, we need only to prove that $\operatorname{Im} R_{A,B} \cap \mathcal{L}_2$

and $\operatorname{Im} R_{A,B}$ are closed simultaneously, and $\dim \mathfrak{L}_2/(\operatorname{Im} R_{A,B} \cap \mathfrak{L}_2) = \operatorname{codim} \operatorname{Im} R_{A,B}$. To this end, introduce a matrix representation

$$R_{A,B} = \begin{bmatrix} R_{11} & 0 \\ R_{21} & R_{22} \end{bmatrix}$$

of the operator $R_{A,B}$, with respect to the decomposition $\mathfrak{L}_1 \dotplus \mathfrak{L}_2$. The block R_{11} of this representation, being a direct sum of scalar operators

$$\chi_+ + \mathcal{F}^{-1} \left(\frac{x-i}{x+i} \right)^{s_j + i \operatorname{Im} \alpha_j} e^{-i\lambda_j x} \mathcal{F} \chi_-$$

with $\lambda_j > 0$ $(j = 1, \dots, m-1)$, is right invertible [8]. Let $R_{11}^{(-1)}$ be one of its right inverses, and put

$$\mathfrak{N} = \left\{ \begin{pmatrix} y \\ R_{21} R_{11}^{(-1)} y \end{pmatrix}, \quad y \in \mathfrak{L}_1 \right\}.$$

Then $\mathfrak{N}$ is closed, $\mathfrak{N} \cap \mathfrak{L}_2 = \{0\}, \mathfrak{N} \dotplus \mathfrak{L}_2 = \mathfrak{L}_1 \dotplus \mathfrak{L}_2$, and $(\mathfrak{L}_2 \cap \operatorname{Im} R_{A,B}) \dotplus \mathfrak{N} = \operatorname{Im} R_{A,B}$. So, $\mathfrak{L}_2 \cap \operatorname{Im} R_{A,B}$ is closed simultaneously with $\operatorname{Im} R_{A,B}$, and $\dim \mathfrak{L}_2/(\operatorname{Im} R_{A,B} \cap \mathfrak{L}_2) = \dim(\mathfrak{L}_2 \dotplus \mathfrak{N})/((\operatorname{Im} R_{A,B} \cap \mathfrak{L}_2) \dotplus \mathfrak{N}) = \operatorname{codim}((\operatorname{Im} R_{A,B} \cap \mathfrak{L}_2) \dotplus \mathfrak{N}) = \operatorname{codim} \operatorname{Im} R_{A,B}$. ∎

According to Lemma 4.1, Fredholm properties of the equation (1.2) are the same as those of the operator $R_{A,B}$ with matrices A and B given by (4.9) and (4.10). Condition (3.10), necessary for (semi-)Fredholmness of $R_{A,B}$, in this case takes the form:

(4.18) $\qquad \operatorname{Re} \alpha_0 = s_0 - s, \ \operatorname{Re} \alpha_m = s_m - s, \ N_0^{-1} \in SAP_p, \ N_m^{-1} \in SAP_p$

On the other hand, the (semi-)Fredholm properties of the operator $R_{A,B}$ are the same as those of W_K, where $K = B^{-1}A$ is now given by:

(4.19) $\qquad K(x) = \begin{bmatrix} f_1(x) & & & & 0 \\ & f_2(x) & & & 0 \\ & & \ddots & & \vdots \\ & & & f_{m-1}(x) & 0 \\ (x-i)^{-\zeta_1} g_1(x) & \cdots\cdots\cdots & & (x-i)^{-\zeta_{m-1}} g_{m-1}(x) & g_m(x) \end{bmatrix}.$

Here f_j, g_j are as in (2.6), and $\zeta_j = s_j - s - \operatorname{Re} \alpha_j \geq 0$ according to (4.8).

Due to (2.3), (4.18) is satisfied automatically for bounded Ω (when $0, m \in \mathcal{J}$), and is equivalent to (2.4) for Ω unbounded. Further,

$$\zeta_j = \begin{cases} 0, & j \in \mathcal{J} \\ \sigma - s - \operatorname{Re} \alpha, & j \in J \end{cases},$$

and hence all $\zeta_j = 0$ whenever condition (2.4) is met. Theorem 2.1 will follow from here as soon as we establish the necessity of condition $\sigma - s - \mathrm{Re}\,\alpha = 0$ for (semi-)Fredholmness of W_K in the case of bounded Ω.

In this case m is even: $m = 2l$. Also, $N_j = I$, $\zeta_j = 0$ for j even, and $N_j = N$, $\zeta_j(=\zeta) = \sigma - s - \mathrm{Re}\,\alpha$ for j odd. Suppose that $\zeta > 0$. Then the local representatives $K_\pm$ of the matrix function (4.19) have the form (3.14) with conditions (3.17), (3.15) and (3.16) satisfied, and $d = 0$. According to Lemma 3.9, $K_\pm$ are APw-factorable, but their partial indices take on both signs. It follows from here and (already proved) "half-line L_p"-versions of Theorems 2.7 and 2.8 that operator W_K is not semi-Fredholm. Therefore, $\sigma - s - \mathrm{Re}\,\alpha = 0$ as soon as W_K is (semi-)Fredholm, regardless of whether or not Ω is bounded.

This completes the proof of Theorem 2.1. ■

4.2. Proof of Theorem 2.10. Let us prove equivalence of *(ii)* and *(iii)*. The matrices $\tilde{K}_\pm$ in our setting are of the form (3.14) with

$$c_j = I, \; d_j = \begin{cases} N_- & \text{if } j \text{ odd} \\ I & \text{if } j \text{ even} \end{cases} \quad \text{for } \tilde{K}_-,$$

$$c_j = \begin{cases} e^{-2\pi i(\sigma + i \mathrm{Im}\,\alpha)}I \\ e^{-2\pi i s}I \end{cases}, \; d_j = \begin{cases} e^{-2\pi i s}N_+ & \text{if } j \text{ odd} \\ e^{-2\pi i s}I & \text{if } j \text{ even} \end{cases} \quad \text{for } \tilde{K}_+.$$

Therefore, conditions (3.15), (3.16) are satisfied, and $c^{-1}d = N_\pm$ for $G = K_\pm$. According to Lemma 3.9, the matrices $K_\pm$ are APw-factorable, and therefore Theorem 2.6 and Corollary 2.9 are applicable. From Corollary 2.9 it follows the equivalence of *(i)* and *(ii)*. Condition *(i)* of Theorem 2.6 coincides with condition *(a)* of Theorem 2.10, condition *(ii)* is equivalent to *(b)* due to Lemma 3.9, statement *(ii)*.

Furthermore, straightforward computations using (3.19) show that in our setting $d(\tilde{K}_-^{-1})\,d(\tilde{K}_+)$ is similar to a direct sum of l copies of the blocks $e^{-2\pi i(\sigma + i \mathrm{Im}\,\alpha)}N_- N_+^{-1}$ and $e^{-2\pi i s}N_-^{-1}N_+$. Therefore, condition *(iii)* of Theorem 2.6 is equivalent to *(c)*, which completes the proof of equivalence *(ii)* $\Leftrightarrow$ *(iii)*.

Finally, formula (2.14) follows from (2.12), since

$$\det \tilde{K}(x) = \left(\frac{x-i}{x+i}\right)^{-(\sigma + s + i\,\mathrm{Im}\,\alpha)nl} \in C(\overline{\mathbb{R}}),$$

and so $\mathrm{Ind}\,\det \tilde{K} = -(\sigma + s)nl$. ■

5. Invertibility and factorability of local representatives.

5.1. Proof of Theorems 2.2, 2.3. Let us consider Banach algebras $\mathcal{A}$ and $\mathcal{B}$ of operators on $L_p(\mathbb{R})$, generated by $\chi_\pm I$ and W_G^0, where $G \in SAP_p$ for $\mathcal{B}$, and $G \in AP_p$ for $\mathcal{A}$. Denote also by $\mathcal{L}_0 = \mathcal{L}_0(L_p^n(\mathbb{R}))$ the ideal of compact operators acting on $L_p^n(\mathbb{R})$.

LEMMA 5.1. $\mathcal{L}_0 \subset \mathcal{B}$.

Proof. Let us identify functions from $L_\infty(\mathbb{R})$ with multiplication operators generated by them on $L_p(\mathbb{R})$. Under this identification, $L_\infty(\mathbb{R})$ is isometrically imbedded in the space of linear bounded operators on $L_p(\mathbb{R})$. Since $\mathcal{B}$ contains all the shift operators $V_h (= \mathcal{F}^{-1} e^{ixh} \mathcal{F})$, and $\chi_\pm$, it contains the (non-closed) algebra of all step-functions, and therefore its closure in $L_\infty(\mathbb{R})$ — the set of all functions having one-side limits on $\mathbb{R}$. In particular, $\mathcal{B} \supset C(\overline{\mathbb{R}})$, whence

(5.1) $$U^{-1} \mathcal{B} U \supset C(\overline{\mathbb{R}_+}).$$

Here U is an isomorphism of $L_p(\mathbb{R}_+)$ onto $L_p(\mathbb{R})$, given by the formula

$$(Uf)(x) = e^{x/p} f(e^x), \quad x \in \mathbb{R}.$$

On the other hand, the function $\varphi(x) \overset{\text{def}}{=} \cotanh \pi(x + \frac{i}{p}) \in SAP_p$ (see [30]), and $U^{-1} W_\varphi U = S_{\mathbb{R}_+} (= S|L_p(\mathbb{R}_+))$. Therefore, $U^{-1} \mathcal{B} U \ni S_{\mathbb{R}_+}$. It follows from here and (5.1) that $U^{-1} \mathcal{B} U$ contains compressions onto $L_p(\mathbb{R}_+)$ of all the operators of the form

(5.2) $$\sum_k a_k(t) \left(\frac{t-i}{t+i} S - S \frac{t-i}{t+i} \right) b_k(t) I \quad (a_k, b_k \in C(\dot{\mathbb{R}})).$$

Since the latter set is dense in $\mathcal{L}_0$, $U^{-1} \mathcal{B} U \supset \mathcal{L}_0(L_p(\mathbb{R}_+))$, and $\mathcal{B} \supset U \mathcal{L}_0(L_p(\mathbb{R}_+)) U^{-1} = \mathcal{L}_0(L_p(\mathbb{R}))$. ∎

The fact that operators of the form (5.2) are dense in $\mathcal{L}_0(L_p(\mathbb{R})$ has one more consequence important for us.

LEMMA 5.2. *For any operator* $K \in \mathcal{L}_0 \left(L_p^n(\mathbb{R}) \right)$ *and any sequence* $h_n \to \infty$, $s\text{-}\lim e^{-ixh_n} K e^{ixh_n} I = 0$.

Proof. It suffices to consider operators K of the form (5.2). For them the result follows from the fact that $\underset{h_n \to \pm\infty}{\text{s-lim}} e^{-ixh_n} S e^{ixh_n} I = \underset{h_n \to \pm\infty}{\text{s-lim}} \mathcal{F}^{-1} \text{sign}(x + h_n) \mathcal{F} = \pm I$. ∎

Let us put

(5.3) $$\mu_\pm(\chi_+) = \chi_+, \ \mu_\pm(\chi_-) = \chi_-, \text{ and } \mu_\pm(\mathcal{F}^{-1} G \mathcal{F}) = \mathcal{F}^{-1} G_\pm \mathcal{F},$$

where $G_\pm$ are the local representatives $G_\pm$ of G at $\pm\infty$. According to Lemma 3.1, statement iv), $G_\pm \in AP_p$ for all $G \in SAP_p$, and therefore the mappings $\mu_\pm$ are defined by (5.3) on generators of the algebra $\mathcal{B}$ and map them onto generators of $\mathcal{A}$.

LEMMA 5.3. *The mappings $\mu_\pm\colon W \mapsto W_\pm$ admit a continuation from (5.3) to the homomorphisms of $\mathcal{B}$ onto $\mathcal{A}$, and*

$$(5.4) \qquad\qquad \forall\, W \in \mathcal{B} \quad \|W_\pm\| \leq |W|.$$

Here $|X|$ stands for the essential norm of the operator X: $|X| = \inf\{\|W + K\|\colon K \in \mathcal{L}_0\}$.

Proof. Consider a set Y of (matrix) functions of the form (3.2) with $a_\pm$ being almost periodic polynomials, a_0 being a piecewise constant function with a compact support, and $u_\pm$ equal to zero in a neighborhood of $\mp\infty$. Of course, Y is dense in SAP_p, and therefore the set

$$W_{ij} \in \{\chi_\pm, W_G\colon G \in Y\}\}$$

is dense in $\mathcal{B}$.

Let us show that for any $W = \sum_i \prod_j W_{ij} \in \mathcal{B}_0$ and a sequence $h_n \to \pm\infty$ there exists a subsequence h_n' such that

$$(5.5) \qquad \operatorname*{s-lim}_{n\to\infty} e^{-ixh_n'} W e^{ixh_n'} = W_\pm \overset{\text{def}}{=} \sum_i \prod_j (W_{ij})_\pm,$$

where $(W_{ij})_\pm$ are given by (5.3).

It suffices to prove (5.5) in case of $W = \mathcal{F}^{-1}G\mathcal{F}$ with a scalar $G \in Y$. Suppose, for the sake of definiteness, $h_n \to +\infty$.

Since for an almost periodic polynomial G_+ the sets of functions $\{G_+(x + h)\}$ and $\{\frac{dG_+}{dx}(x + h)\}$, $h \in \mathbb{R}$, are relatively compact, there exists a subsequence h_n' of the sequence $h_n \to +\infty$ such that

$$(5.6) \qquad (\forall x \in \mathbb{R}) \ \|G_+(x) - G_+(x + h_n')\| < \frac{1}{n}, \ \left\|\frac{dG_+}{dx}(x) - \frac{dG_+}{dx}(x + h_n')\right\| < \frac{1}{n}.$$

According to [35, Theorem 3.2.2], for any $f \in L_p(\mathbb{R})$ and $\varphi \in L_1(\mathbb{R})$ such that $\int_{\mathbb{R}} \varphi(x)\, dx = 1$,

$$(5.7) \qquad\qquad \lim_{\varepsilon\to 0} \|f * \varphi_\varepsilon - f\|_{L_p(\mathbb{R})} = 0,$$

where $\varphi_\varepsilon(x) = \varepsilon^{-1}\varphi(x/\varepsilon)$, $\varepsilon > 0$. Letting φ in (5.7) to be a rapidly decreasing function in S with a compact support of its Fourier transform, we find from here that a set

$$\Phi = \{f \in L_p(\mathbb{R})\colon \mathcal{F}f \text{ has compact support }\}$$

is dense in $L_p(\mathbb{R})$. Therefore, for $G \in Y$, $f \in \Phi$ and h_n' big enough:

$$(5.8) \quad e^{-ixh_n'}\mathcal{F}^{-1}G\mathcal{F}e^{ixh_n'}f - \mathcal{F}^{-1}G_+\mathcal{F}f =$$
$$\mathcal{F}^{-1}\big(G_+(x + h_n')u_+(x + h_n') - G_+(x) + G_-(x + h_n')(1 - u_+(x + h_n')) + G_0(x + h_n')\big)\mathcal{F}f =$$
$$\mathcal{F}^{-1}\big(G_+(x + h_n') - G_+(x)\big)\tilde{\chi}(x)\mathcal{F}f,$$

where $\tilde{\chi} \in C_0^\infty(\mathbb{R})$ and $\tilde{\chi} \equiv 1$ on supp $\mathcal{F}f$.

According to Mikhlin's theorem on multiplicators [23],

(5.9)
$$\|\mathcal{F}^{-1}(G_+(x + h'_n) - G_+(x))\tilde{\chi}(x)\mathcal{F}\| \leq \text{const } \frac{1}{n}.$$

Indeed, from (5.6) and compactness of supp $\tilde{\chi}$ it follows that

$$|(G_+(x + h'_n) - G_+(x))\tilde{\chi}(x)| \leq \frac{1}{n}\|\tilde{\chi}\|_{C(\mathbb{R})},$$

$$|x|\left|\left(\frac{dG_+}{dx}(x + h'_n) - \frac{dG_+}{dx}(x)\right)\tilde{\chi}(x) + (G_+(x + h'_n) - G_+(x))\frac{d\tilde{\chi}}{dx}(x)\right| \leq c\frac{1}{n}$$

for all $x \in \mathbb{R}$, and $c = \sup\{|x|(|\tilde{\chi}(x)| + |\tilde{\chi}'(x)|) : x \in \mathbb{R}\}$. From (5.8)–(5.9) it follows that (5.5) holds for $W = \mathcal{F}^{-1}G\mathcal{F}$, and therefore, for all $W \in \mathcal{B}^0$.

According to (5.5) and Lemma 5.2, for all $W \in \mathcal{B}^0$ and $K \in \mathcal{L}_0$:

$$W_\pm = \text{s-lim}_{n\to\infty} e^{-ixh'_n}(W + K)e^{ixh'_n}I,$$

which implies (5.4) for $W \in \mathcal{B}^0$, and hence for all $W \in \mathcal{B}$. ∎

THEOREM 5.4. *If operator $W \in \mathcal{B}$ is Fredholm in $L_p(\mathbb{R})$, then the corresponding operators $\mu_\pm(W) = W_\pm \ (\in \mathcal{A})$ are invertible in this space.*

Proof. Let ind $W = m$. Consider the operator $W' = WW_D$, where

$$D(t) = \text{diag}[\left(\frac{t + i}{t - i}\right)^m, 1, \ldots, 1].$$

Since $D_\pm = I$, it follows from (5.3) that $\mu_\pm(W_D) = I$, and therefore

(5.10)
$$\mu_\pm(W') = \mu_\pm(W) = W_\pm.$$

According to [7, Theorem 3.2], the operator W_D is Fredholm, and ind $W_D = -m$. Hence, W' is a Fredholm operator with index 0, and, as any such operator, it can be represented in a form

(5.11)
$$W' = V + K$$

where operator V is invertible and $K \in \mathcal{L}_0$.

Due to Lemma 5.1, $K \in \mathcal{B}$. Since $W, W_D \in \mathcal{B}$, it follows from here and (5.11) that $V \in \mathcal{B}$. Applying Lemmas 5.2 and 5.3, we find that for any sequence $h_n \to \pm\infty$ there exists a subsequence h'_n such that

$$W_\pm = V_\pm = \text{s-lim } e^{-ixh'_n}Ve^{ixh'_n}I.$$

Taking adjoints in (5.11), and then applying the same reasoning to the invertible operator $V^* \in \mathcal{B}$ and the sequence h'_n, we find its subsequence h''_n such that

$$W^*_\pm = V^*_\pm = \text{s-lim}\, e^{-ixh''_n} V^* e^{ixh''_n} I.$$

Of course, without loss of generality we may suppose that $h'_n = h''_n\, (= l_n)$.

The operators $e^{-ixl_n} V e^{ixl_n} I$, $e^{-ixl_n} V^* e^{ixl_n} I$ are invertible, and norms of their inverses

$$\|(e^{-ixl_n} V e^{ixl_n} I)^{-1}\| = \|(e^{-ixl_n} V^* e^{ixl_n} I)^{-1}\| =$$
$$\|e^{-ixl_n} V^{-1} e^{ixl_n} I\| = \|e^{-ixl_n} V^{*-1} e^{ixl_n} I\| = \|V^{-1}\|$$

are bounded above. According to [10, Lemma 3.1.1], it means that the s-limits $W_\pm$ of the sequences of these operators are invertible. ∎

Theorem 2.3 is a particular case of Theorem 5.4, since for any $G \in SAP_p$ $R_{G,I} \in \mathcal{B}$, and $\mu_\pm(R_{G,I}) = R_{G_\pm,I}$.

If operator W_G is n- $(d$-$)$ normal in the space $L_2(\mathbb{R})$, then $W^*_G W_G$ (respectively, $W_G W^*_G$) is Fredholm. According to Theorem 5.4, the operators $W^*_{G_\pm} W_{G_\pm}$ (respectively, $W_{G_\pm} W^*_{G_\pm}$) are invertible. Then, of course, $W_{G_\pm}$ are left (right) invertible. This takes care of Theorem 2.2.

5.2. Proof of Theorem 2.5. Denote by B_2 the Hilbert space of Bezikovich almost periodic functions, that is, a completion of the set of almost periodic polynomials $f(t) = \sum_j a_j e^{i\lambda_j t}$ with respect to the norm

$$\|f\|_{B_2} = \left(\sum_j |a_j|^2\right)^{1/2} = \left(\mathbf{M}(|f|)^2\right)^{1/2}.$$

It is well known that $f \in B_2$ admits a unique continuation to a function $\tilde{f} \in L_2(\mathbb{R}_B)$, where $\mathbb{R}_B$ is the Bohr compact of maximal ideals of AP,

$$\|\tilde{f}\|_{L_2(\mathbb{R}_B)} = \left(\int_{\mathbb{R}_B} |\tilde{f}(t)|^2\, d\mu\right)^{1/2},$$

and μ is a Haar invariant measure on $\mathbb{R}_B$ (see [26]). In what follows we will not distinguish between $f \in B_2$ and $\tilde{f} \in L_2(\mathbb{R}_B)$.

Let us define orthoprojections $\mathcal{P}_\pm$ and $\tilde{\mathcal{P}}_\pm$ in B_2 according to the formulas:

$$\mathcal{P}_+(e^{i\lambda t}) = \begin{cases} e^{i\lambda x} & \text{if } \lambda \geq 0 \\ 0 & \text{otherwise} \end{cases}, \quad \tilde{\mathcal{P}}_+(e^{i\lambda t}) = \begin{cases} e^{i\lambda x} & \text{if } \lambda > 0 \\ 0 & \text{otherwise} \end{cases},$$

$$\mathcal{P}_- = I - \mathcal{P}_+, \tilde{\mathcal{P}}_- = I - \tilde{\mathcal{P}}_+.$$

LEMMA 5.5. (see [14, Theorem 3.5]) *Let G be $n \times n$ matrix function in AP. Then the operators $T_G = P_+ + GP_-$ on $L_2(\mathbb{R})$, and $\mathcal{T}_G = \mathcal{P}_+ + G\mathcal{P}_-$, $\tilde{\mathcal{T}}_G = \tilde{\mathcal{P}}_+ + G\tilde{\mathcal{P}}_-$ on B_2 are simultaneously invertible.*

Let us mention that coincidence of spectra of other naturally related operators on $L_2(\mathbb{R})$ and B_2 was established in [1, 29].

We also need the following result of Kurbatov [19, Section2.2]

LEMMA 5.6. *For $G \in AP_W$ the operator W_G is simultaneously invertible in the spaces $L_p(\mathbb{R}_+)$, $1 < p < \infty$.*

Since the operators W_G on $L_2(\mathbb{R}_+)$ and T_G on $L_2(\mathbb{R})$ have the same invertibility properties, due to Lemmas 5.5 and 5.6 we are left with a proof of the following.

LEMMA 5.7. *Let $G \in AP_W$ and the corresponding operator T_G be invertible on B_2. Then $G = G^+ G^-$, where $G^\pm, (G^\pm)^{-1} \in AP_W^\pm$.*

Proof. The operator $C T_G^* C = \tilde{\mathcal{P}}_- + \tilde{\mathcal{P}}_+ G^T I$ is invertible simultaneously with T_G, where C stands for complex conjugation. In its turn, invertibility of $\tilde{\mathcal{P}}_- + \tilde{\mathcal{P}}_+ G^T I$ implies invertibility of $\tilde{\mathcal{P}}_- + G^T \tilde{\mathcal{P}}_+$. Since the matrix function G itself is invertible due to invertibility of T_G (see [4, 22]), this means that operators $T'_G = G^{-1}\mathcal{P}_+ + \mathcal{P}_-$ and $T''_G = (G^{-1})^T \tilde{\mathcal{P}}_- + \tilde{\mathcal{P}}_+$ are also invertible. Put

$$(5.12) \qquad \Phi^\pm = \mathcal{P}^\pm (T'_G)^{-1} I_n, \quad \Psi^\pm = \tilde{\mathcal{P}}^\pm (T''_G)^{-1} I_n,$$

(the operators $\mathcal{P}_\pm$ are applied to $n \times n$ matrices column-wise). Then, of course, $\Phi^\pm, \Psi^\pm \in B_2$. However, stronger inclusions hold. Namely,

$$(5.13) \qquad \Phi^+, \Psi^+ \in AP_W^+, \quad \Phi^-, \Psi^- \in AP_W^-, \text{ and } M(\Phi^-) = M(\Psi^+) = 0.$$

To prove this, let us introduce a C^*-algebra $\mathcal{R}$ of operators on B_2, generated by multiplications by constant matrices, and operators of the form $e^{i\lambda t}\mathcal{P}_+ e^{-i\lambda t}$, $\lambda \in \mathbb{R}$, and put

$$\Upsilon_W = \{\sum_\lambda A_\lambda e^{i\lambda t} I : A_\lambda \in \mathcal{R} \text{ and } \sum_\lambda \|A_\lambda\| < \infty\}.$$

According to [2, Proposition 3], the algebra Υ_W is inverse closed, that is, $X^{-1} \in \Upsilon_W$ whenever $X \in \Upsilon_W$ and X is invertible. In particular, $(T'_G)^{-1} \in \Upsilon_W$, i.e. $(T'_G)^{-1} = \sum_\lambda B_\lambda e^{i\lambda t} I$, where $B_\lambda \in \mathcal{R}$ and $\sum \|B_\lambda\| < \infty$. Hence,

$$(T'_G)^{-1}(I) = \sum_\lambda B_\lambda(e^{i\lambda t} I) = \sum_\lambda C_\lambda e^{i\lambda t} \in AP_W,$$

since for constant matrices $C_\lambda = e^{-i\lambda t} B_\lambda(e^{i\lambda t} I)$ $\|C_\lambda\| \leq \|B_\lambda\|$.

From here and (5.12) follow the properties of Φ stated in (5.13). The corresponding properties of Ψ can be proved analogously.

Directly from the definition (5.12) it follows that

(5.14) $$\Phi^+ = G(I_n - \Phi^-), \quad (\Psi^-)^T = (I_n - \Psi^+)^T G,$$

and therefore

(5.15) $$C \stackrel{\text{def}}{=} (I_n - \Psi^+)^T G(I_n - \Phi^-) = (I_n - \Psi^+)^T \Phi^+ = (\Psi^-)^T(I_n - \Phi^-)$$

belongs to the classes AP_W^+ and AP_W^- simultaneously. In other words, $C \in \mathbb{C}^{n \times n}$.

Let us show now that C is nonsingular. To do that, we follow the lines of the proof of [22, Theorem 3.4]. Namely, suppose that $\det C = 0$. Then, according to (5.15), $\det \Psi^- \cdot \det(I - \Phi^-) \equiv 0$ in the lower half-plane Π^-. Since $\mathbf{M}(\Phi^-) = 0$, $\lim_{y \to -\infty} \det(I - \Phi^-(iy)) = 1$, and therefore $\det \Psi^- \equiv 0$ (in the neighborhood of $-\infty$ in Π^-, and thus on the whole Π^-, in particular, on $\mathbb{R}$). From here and (5.14) it follows that $\det(I - \Psi^+(t)) = 0$ for $t \in \mathbb{R}$. So, $\det(I - \Psi^+) \equiv 0$ in the upper half-plane Π^+. In particular, $\det(I - \Psi^+(iy)) = 0$ for all $y > 0$, which contradicts the condition $\mathbf{M}(\Psi^+) = 0$. Thus, $\det C \neq 0$.

Now put $G^+ = \Phi^+$ and $G^- = C^{-1}(\Psi^-)^T$. Then, according to (5.15),

$$(G^+)^{-1} = C^{-1}(I_n - \Psi^+)^T, \quad (G^-)^{-1} = I_n - \Phi^-.$$

It follows from here that $(G^+)^{\pm 1} \in AP_W^+, (G^-)^{\pm 1} \in AP_W^-$. Finally, use (5.15) to conclude $G^+ G^- = \Phi^+ C^{-1}(\Psi^-)^T = \Phi^+ C^{-1} C C^{-1}(\Psi^-)^T = (I - \Psi^+)^{T-1} C(I - \Phi^-)^{-1} = G.$ ∎

REFERENCES

[1] A. B. Antonevich, *Linear functional equations. Operator approach*, University Press, Minsk, 1988 (in Russian).

[2] R. G. Babadzhanyan and V. S. Rabinovich, *On factorization of almost periodic matrix functions*, Differential and Integral Equations, and Complex Analysis (Elista), University Press, Elista, 1986, pp. 13–22 (in Russian).

[3] M. A. Bastos and A. F. dos Santos, *Wiener-Hopf operators with oscillating symbols and convolution operators on a union of intervals*, Integral Equations and Operator Theory 15 (1992), 920–941.

[4] K. F. Clancey and I. Gohberg, *Factorization of matrix functions and singular integral operators*, Birkhäuser, Basel and Boston, 1981.

[5] L. Coburn and R. G. Douglas, *Translation operators on the half-line*, Proc. Nat. Acad. Sci. USA 62 (1969), 1010–1013.

[6] R. V. Duduchava, *On convolution type integral operators with discontinuous coefficients*, Math. Nachr. 79 (1977), no. 1, 75–98.

[7] ———, *Convolution type integral equations with discontinuous presymbols, singular integral equations with fixed singularities, and their applications to problems in mechanics*, Trudy Tbiliss. Matem. Inst. Razmadze 60 (1979), 5–135 (in Russian), English translation: *Integral equations with fixed singularities*. Teubner, Leipzig, 1979.

[8] R. V. Duduchava and A. I. Saginashvili, *Convolution integral equations on a half-line with semi-almost-periodic presymbols*, Differentsial'nye Uravneniya 17 (1981), 301–312 (in Russian), English translation in *Differential Equations*.

[9] I. Gohberg and I. Feldman, *Integro-difference Wiener-Hopf equations*, Acta Sci. Math. Szeged **30** (1969), no. 3–4, 199–224 (in Russian).

[10] _____, *Convolution equations and projection methods for their solution*, Nauka, Moscow, 1971 (in Russian), English translation *Amer. Math. Soc. Transl. of Math. Monographs* **41**, Providence, R.I. 1974.

[11] I. Gohberg and M. G. Krein, *Systems of integral equations on a half-line with kernel depending upon the difference of the arguments*, Uspehi Mat. Nauk **13** (1958), no. 2, 3–72 (in Russian), English translation in *Amer. Math. Soc. Transl.* **14** (1960), no. 2, 217–287.

[12] L. Hörmander, *Estimates for translation invariant operators in L^p spaces*, Acta Math **104** (1960), 93–140.

[13] E. Jahnke and F. Emde, *Tables of functions with formulae and curves*, Dover, NY, 1943.

[14] Yu. I. Karlovich and G. S. Litvinchuk, *On some classes of semi-Fredholm operators*, Izvestiya VUZov. Mat. (1990), no. 2, 3–16 (in Russian).

[15] Yu. I. Karlovich and I. M. Spitkovsky, *Factorization of almost periodic matrix functions and Fredholm theory of Toeplitz operators with semi almost periodic matrix symbols*, Linear and Complex Analysis Problem Book: 199 research problems, Lecture Notes in Mathematics **1043** (1984), 279–282.

[16] _____, *Factorization of almost periodic matrix-valued functions and the Noether theory for certain classes of equations of convolution type*, Izv. Akad. Nauk SSSR, Ser. Mat **53** (1989), no. 2, 276–308 (in Russian), English translation in Mathematics of the USSR, Izvestiya **34** (1990), 281–316.

[17] S. G. Krein, *Linear equations in a Banach space*, Birkhäuser, Basel-Boston, 1982.

[18] N. Ya. Krupnik, *Banach algebras with symbol and singular integral operators*, Birkhäuser, Basel-Boston, 1987.

[19] V. G. Kurbatov, *Linear differential-difference equations*, University Press, Voronezh, 1990 (in Russian).

[20] B. M. Levitan, *Almost periodic functions*, GITTL, Moscow, 1953 (in Russian).

[21] J. L. Lions and E. Magenes, *Problémes aux limites non homogénes et applications*, Dunod, Paris, 1968–1970.

[22] G. S. Litvinchuk and I. M. Spitkovsky, *Factorization of measurable matrix functions*, Birkhäuser Verlag, Basel and Boston, 1987.

[23] S. G. Mikhlin, *Multidimensional singular integrals and integral equations*, Phizmatgiz, Moscow, 1962 (in Russian).

[24] B. V. Pal'cev, *Convolution equations on a finite interval for a class of symbols having power asymptotics at infinity*, Izv. Akad. Nauk SSSR. Mat. **44** (1980), 322–394 (in Russian), English translation in *Math. USSR Izv.* **16** (1981).

[25] _____, *A generalization of the Wiener-Hopf method for convolution equations on a finite interval with symbols having power asymptotics at infinity*, Mat. Sb. **113** (**155**) (1980), 355–399 (in Russian), English translation in *Math. USSR Sb.* **41** (1982).

[26] A. A. Pankov, *Bounded and almost periodic solutions of nonlinear differential operator equations*, Naukova Dumka, Kiev, 1985 (in Russian).

[27] A. I. Saginashvili, *Singular integral equations with coefficients having discontinuities of semi-almost periodic type*, Trudy Tbiliss. Mat. Inst. Razmadze **66** (1980), 84–95 (in Russian), English translation: Amer. Math. Soc. Transl. **127**, no. 2 (1986).

[28] D. Sarason, *Toeplitz operators with semi-almost periodic symbols*, Duke Math. J. **44** (1977), no. 2, 357–364.

[29] M. A. Shubin, *On the coincidence of spectrums of pseudodifferential almost periodic operators on $L_2(\mathbb{R}^n)$ and $B_2(\mathbb{R}^n)$*, Sibirsk. Mat. Zh. **17** (1976), no. 4, 200–215 (in Russian).

[30] I. B. Simonenko and Chin Ngok Min, *Local approach to the theory of one-dimensional singular integral equations with piecewise continuous coefficients. Noetherity*, University Press, Rostov on Don, 1986 (in Russian).

[31] I. M. Spitkovsky, *Factorization of certain classes of semi-almost-periodic matrix functions and its applications to systems of equations of convolution type*, Izvestiya VUZ., Mat. (1983), no. 4, 88–94 (in Russian), English translation in *Soviet Math. – Iz. VUZ* **27** (1983), 383–388.

[32] I. M. Spitkovsky and P. M. Tishin, *Factorization of new classes of almost-periodic matrix functions*, Reports of the extended sessions of a seminar of the I. N. Vekua Institute for Applied Mathematics **3** (1989), no. 1, 170–173 (in Russian).

[33] I. M. Spitkovsky and S. I. Yatsko, *On the Riemann boundary value problem with matrix coefficients admitting infinite partial indices*, Izvestiya VUZov. Mat (1985), no. 6, 45–53 (in Russian), English translation in *Soviet Mathematics (Iz. VUZ)* **29** (1985), 55–65.

[34] S. B. Stechkin, *On bilinear forms*, Dokl. AN SSSR **71** (1950), 237–240 (in Russian).

[35] E. Stein, *Singular integrals and differentiability properties of functions*, Princeton University Press, Princeton, NJ, 1970.

[36] H. Triebel, *Interpolation theory, function spaces, differential operators*, North-Holland, Amsterdam, 1978.

[37] H. Widom, *Singular integral equations in L_p*, Trans. Amer. Math. Soc. **97** (1960), 131–160.

Hydroacoustic Department   Department of Mathematics
Marine Hydrophysical Institute   The College of William and Mary
Ukrainian Academy if Science   Williamsburg, Virginia 23187–8795
Street of the Soviet Army 3   USA
270100 Odessa, Ukraine

MSC 1991: Primary 45E10
            Secondary 42A75, 46F99, 47A53, 47A68

Operator Theory:
Advances and Applications, Vol. 71
© 1994 Birkhäuser Verlag Basel/Switzerland

KERNELS OF TOEPLITZ OPERATORS

Donald Sarason

In honor of Harold Widom and his deep contributions to the study of Toeplitz operators.

A new proof is presented of E. Hayashi's characterization of the kernels of Toeplitz operators on H^2 of the unit disk.

1. Introduction.

Which subspaces of the Hardy space H^2 of the unit disk, D, are kernels of Toeplitz operators? Although the question may at first sound unpromising, it turns out to have an interesting answer, thanks to the work of E. Hayashi [4], [5], [6]. The purpose of this note is to present an account of Hayashi's result, using methods different from his. All notations will be standard. In particular, for φ a function in L^∞ of ∂D, the Toeplitz operator on H^2 with symbol φ will be denoted by T_φ; it is defined by $T_\varphi h = P(\varphi h)$, where P is the orthogonal projection in L^2 (of ∂D) with range H^2. The Toeplitz operator T_z, the unilateral shift operator on H^2, will be denoted by S.

The kernel of a Toeplitz operator is stable under division by inner functions: if h is in the kernel of T_φ and the inner function u divides h, then h/u is in the kernel of T_φ. (Reason: Under the given conditions, $h/u = T_{\bar u}T_\varphi h$.) In particular, if h is in the kernel of T_φ and $h(0) = 0$, then S^*h is in the kernel of T_φ. The subspaces of H^2 with the preceding property, which are called nearly S^*-invariant subspaces, have been characterized by D. Hitt [8]. In [13] the author presented an alternative approach to and a refinement of Hitt's result based on a family of transformations that map H^2 onto certain Hilbert spaces contained contractively in H^2. These same spaces and transformations enable one to give a quick proof of Hayashi's theorem. They are described in Section 2. Hitt's theorem is discussed in Section 3 and Hayashi's in Section 4.

      Hayashi's theorem reveals a close connection between kernels of Toeplitz operators and rigid functions in H^1. A function in H^1 is called rigid if no other functions in H^1, except for positive scalar multiples of itself, have the same argument as it almost everywhere on ∂D. One can obtain a preview of the connection by considering one-dimensional kernels. Suppose the kernel of the Toeplitz operator T_φ is the one-dimensional subspace spanned by the function f. By the stability property mentioned in the preceding paragraph, the function f must then be an outer function. Since $T_\varphi f = 0$, the function φf is in $(H^2)^\perp$, so $\varphi f = \bar{z}\bar{g}$ with g in H^2, say $g = ug_0$ with u inner and g_0 outer. But then $T_\varphi u f = 0$, so uf is a constant multiple of f, in other words, u is constant, and g is an outer function. The function g/f is thus an outer function in H^∞. We have the factorization

$$T_\varphi = T_{\bar{g}/\bar{f}} T_{\bar{z}\bar{f}/f}.$$

The kernel of the left factor on the right side is trivial, because g/f is an outer function. Thus, the kernel of the operator $T_{\bar{z}\bar{f}/f}$ is the one-dimensional subspace spanned by f. It is asserted that this forces the function f^2 to be rigid. Indeed, suppose the function G in H^1 has the same argument as f^2 almost everywhere on ∂D, but is not a positive multiple of f^2. We can assume G is an outer function, for if its inner factor, say v, is nonconstant, we can replace G either by $(1+v)^2 G/v$ or by $-(1-v)^2 G/v$; those two outer functions will have the same argument as f^2 almost everywhere on ∂D, and at least one of them will fail to be a positive multiple of f^2. Assuming then that G is outer, we consider the H^2 function $f_1 = G^{1/2}$, which is not a scalar multiple of f, but which satisfies $\bar{f}_1/f_1 = \bar{f}/f$ almost everywhere on ∂D. From the last equality one sees that $T_{\bar{z}\bar{f}/f} f_1 = 0$, in contradiction to the one-dimensionality of the kernel of $T_{\bar{z}\bar{f}/f}$. Conclusion: If a one-dimensional subspace of H^2 is the kernel of a Toeplitz operator, it is spanned by a function whose square is rigid.

      The converse of the preceding conclusion is true. In fact, suppose f is a function in H^2 whose square is rigid. It is asserted that the kernel of the operator $T_{\bar{z}\bar{f}/f}$ is the one-dimensional subspace spanned by f. Indeed, it is clear that f belongs to that kernel. Suppose the nonzero function g also belongs, say $g = ug_0$ with u inner and g_0

outer. Then $\bar{z}\bar{f}ug_0/f$ is in $(H^2)^\perp$ and has the same modulus on ∂D as g_0, so it equals $\bar{z}\bar{v}\bar{g}_0$ for some inner function v. That means $\bar{f}/f = \bar{u}\bar{v}\bar{g}_0/g_0$, in other words, the functions f^2 and uvg_0^2 have the same argument almost everywhere on ∂D. Thus uvg_0^2 is a positive multiple of f^2. But uvg_0^2 and $(1+uv)^2g_0^2$ also have the same argument almost everywhere on ∂D and so are positive multiples of each other, which forces uv, and hence both u and v, to be constant. The upshot is that g is an outer function, and g^2 is a positive multiple of f^2, implying that g is a constant multiple of f, as desired. Conclusion: A one-dimensional subspace of H^2 is the kernel of a Toeplitz operator if and only if it is spanned by a function whose square is rigid.

Although considerable effort has been expended in trying to understand rigid functions, their structure remains rather mysterious. A brief discussion is contained in the concluding Section 5. To close this section, the following characterization [1], a slight variant of the one at the end of the last paragraph, is mentioned. (Note that, by reasoning used above, rigid functions are outer functions, so every rigid function is the square of an outer function in H^2.)

Proposition 1. If f is an outer function in H^2, then f^2 is rigid if and only if the operator $T_{\bar{f}/f}$ has a trivial kernel.

In fact, we have the factorization $T_{\bar{z}\bar{f}/f} = S^*T_{\bar{f}/f}$. The kernel of S^* is the one-dimensional subspace of constant functions. The operator $T_{\bar{f}/f}$ maps the function f into that subspace. Hence the kernel of $T_{\bar{f}/f}$ is trivial if and only if the kernel of $T_{\bar{z}\bar{f}/f}$ is one-dimensional. As shown above, the latter happens if and only if f^2 is rigid.

2. De Branges–Rovnyak Spaces.

The Hilbert spaces that are useful in connection with the theorems of Hitt and Hayashi were originally introduced by L. de Branges and J. Rovnyak [2]. The starting point is a function b in the unit ball of H^∞. The de Branges–Rovnyak space $\mathcal{H}(b)$ consists by definition of the range of the operator $(1 - T_bT_{\bar{b}})^{1/2}$, with the range norm (that is, the norm making $(1 - T_bT_{\bar{b}})^{1/2}$ a coisometry of H^2 onto $\mathcal{H}(b)$). The needed properties of these spaces are in the literature and will be stated here with references (usually not the original references). The following two properties can be found, for example, in [11].

Proposition 2. *The kernel function in $\mathcal{H}(b)$ for the evaluation functional at the point w of D is the function $k_w^b = (1 - \overline{b(w)}b)k_w$, where $k_w(z) = (1 - \bar{w}z)^{-1}$ (the kernel function in H^2 for evaluation at w).*

Proposition 3. *The space $\mathcal{H}(b)$ is invariant under the operator S^*, and S^* acts as a contraction in it.*

Note that if u is an inner function, the space $\mathcal{H}(u)$ is an ordinary subspace of H^2, namely, it is the orthogonal complement in H^2 of the subspace uH^2. By Beurling's theorem, therefore, the spaces $\mathcal{H}(u)$ with u inner are the proper S^*-invariant subspaces of H^2. (Hitt's theorem associates an S^*-invariant subspace with each nearly S^*-invariant subspace.)

The decomposition in the next propopsition is established, among other places, in [10].

Proposition 4. *If u is an inner function, then the space $\mathcal{H}(ub)$ is the orthogonal direct sum of the subspaces $\mathcal{H}(u)$ and $u\mathcal{H}(b)$. The inclusion map of $\mathcal{H}(u)$ into $\mathcal{H}(ub)$ is an isometry, and the operator T_u acts as an isometry from $\mathcal{H}(b)$ into $\mathcal{H}(ub)$.*

Of primary concern here will be functions b that arise in the following manner. Assume given an outer function f in H^2. We form the Herglotz integral of $|f|^2$, that is, the function F in D given by

$$F(z) = \frac{1}{2\pi} \int_{\partial D} \frac{e^{i\theta} + z}{e^{i\theta} - z} |f(e^{i\theta})|^2 d\theta.$$

This function has a positive real part, so it can be written as $\frac{1+b}{1-b}$ with b in the unit ball of H^∞; explicitly, $b = \frac{F-1}{F+1}$. We also define the function $a = \frac{2f}{F+1}$, so that $f = \frac{a}{1-b}$. The function a is an outer function, and a simple calculation shows that $|a|^2 + |b|^2 = 1$ almost everywhere on ∂D. We shall call (b, a) the pair associated with f.

The functions b that arise in the preceding manner have the special property that $1 - |b|^2$ is log-integrable on ∂D. Suppose we are given a function b satisfying the last condition. Let a be an outer function whose modulus on ∂D equals $(1 - |b|^2)^{1/2}$ almost everywhere. (It is unique to within a unimodular multiplicative constant.) The function $\frac{1+b}{1-b}$ has a positive real part and so is, to within addition of an imaginary constant, the

Herglotz integral of a positive measure μ on ∂D:

$$\frac{1 + b(z)}{1 - b(z)} = ic + \int_{\partial D} \frac{e^{i\theta} + z}{e^{i\theta} - z} d\mu(e^{i\theta}).$$

Since $\operatorname{Re}\left(\frac{1+b}{1-b}\right) = \frac{1-|b|^2}{|1-b|^2}$, the function $\left|\frac{a}{1-b}\right|^2$ is the Radon–Nikodym derivative of the absolutely continuous component of μ with respect to normalized Lebesgue measure. In particular, the function $f = \frac{a}{1-b}$, which is outer, belongs to H^2. We shall call (b, a) an admissible pair, and if μ is absolutely continuous we shall say the pair is special.

In [14] it is shown that there is an isometry V of $H^2(\mu)$ (the closure of the polynomials in $L^2(\mu)$) onto $\mathcal{H}(b)$ given by

$$(Vq)(z) = (1 - b(z)) \int_{\partial D} \frac{q(e^{i\theta})}{1 - e^{-i\theta} z} d\mu(e^{i\theta}).$$

The proof is basically a computation resulting in the equality $Vk_w = (1 - \overline{b(w)})^{-1} k_w^b$ ($w \in D$). If one precedes V by the isometry of H^2 into $H^2(\mu)$ of multiplication by $1/f$ (whose range is $H^2(|f|^2 \frac{d\theta}{2\pi})$), one obtains an isometry of H^2 into $\mathcal{H}(b)$; its range is all of $\mathcal{H}(b)$ if and only if the pair (b, a) is special. The latter map is easily seen to equal $T_{1-b} T_{\bar{f}}$. These conclusions are summarized by the next proposition.

Proposition 5. *Let (b, a) be an admissible pair, and let $f = \frac{a}{1-b}$. Then the operator $T_{1-b} T_{\bar{f}}$ is an isometry of H^2 into $\mathcal{H}(b)$. Its range is all of $\mathcal{H}(b)$ if and only if the pair is special.*

The following consequence is explained in [13].

Corollary. *Let (b, a) be a special admissible pair and let $f = \frac{a}{1-b}$. Then $T_{1-b} T_{\bar{f}} T_f T_{1-\bar{b}} = 1 - T_b T_{\bar{b}}$.*

The following rigidity criterion is from [14].

Proposition 6. *Let (b, a) be an admissible pair and let $f = \frac{a}{1-b}$. Then aH^2 is contained in $\mathcal{H}(b)$. If (b, a) is special, then f^2 is rigid if and only if aH^2 is dense in $\mathcal{H}(b)$. If aH^2 is dense in $\mathcal{H}(b)$ then (b, a) is special.*

The proof will be sketched. Standard properties of Toeplitz operators yield the equality $T_{1-b} T_{\bar{f}} T_{f/\bar{f}} = T_a$, implying that the operator $T_{1-b} T_{\bar{f}}$ maps the range of the operator $T_{f/\bar{f}}$ onto aH^2. Therefore, by Proposition 5, aH^2 is contained in $\mathcal{H}(b)$. If (b, a) is special we can conclude, again using Proposition 5, that aH^2 is dense in $\mathcal{H}(b)$ if and only if the operator $T_{f/\bar{f}}$ has a dense range. By Proposition 1, the latter happens if and only if f^2 is rigid. Finally, if aH^2 is dense in $\mathcal{H}(b)$, then the range of $T_{1-b}T_{\bar{f}}$ is dense in $\mathcal{H}(b)$, which implies that (b, a) is special, once more by Proposition 5.

Corollary. *Let (b, a) be a special admissible pair and let $f = \frac{a}{1-b}$. If f^2 is rigid and u is any inner function, then (ub, a) is special and $\left(\frac{a}{1-ub}\right)^2$ is rigid.*

This is contained in [14], but the following argument is more direct than the one there. First, from the inequality $1 - T_b T_{\bar{b}} \leq 1 - T_{ub}T_{\overline{ub}}$ one can conclude that $\mathcal{H}(b)$ is contained contractively in $\mathcal{H}(ub)$. (This depends upon a well-known criterion of R. G. Douglas [3].) Moreover $\mathcal{H}(b)$ is dense in $\mathcal{H}(ub)$ because $\mathcal{H}(b)$ contains the polynomials and the polynomials are dense in $\mathcal{H}(ub)$ [12]. The rest now follows by Proposition 6.

An example in [14] shows that the converse of the last corollary fails.

3. Hitt's Theorem.

If M is a nontrivial nearly S^*-invariant subspace of H^2, then M is not contained in H_0^2, so the orthogonal complement in M of $M \cap H_0^2$ has dimension one.

Hitt's Theorem. *Let M be a nontrivial nearly S^*-invariant subspace of H^2, and let g be the function of unit norm in M that is orthogonal to $M \cap H_0^2$ and positive at the origin. If h is any function in M, then the quotient h/g is in H^2 and has the same norm as h. The subspace $M' = \{h/g : h \in M\}$ is S^*-invariant.*

Thus, according to the theorem, $M = T_g M'$, where M' is an S^*-invariant subspace on which T_g acts isometrically.

The proof will be sketched. Full details can be found in [13].

One easily verifies that the function $g(0)g$ is the kernel function in M for the evaluation functional at the origin: if h is in M, then $\langle h, g(0)g \rangle = h(0)$. Now if h is in M then the function $h - h(0)g(0)^{-1}g$, which in view of the preceding observation can be written as $h - (g \otimes g)h$, is also in M, and it vanishes at the origin, so its image under S^* belongs to M. Thus M is invariant under the operator $R = S^*(1 - g \otimes g)$.

Fix h in M and let $c_0 = \langle h, g \rangle$. Then

$$h - c_0 g = (1 - g \otimes g)h = SS^*(1 - g \otimes g)h$$

$$= SRh,$$

so that

$$h = c_0 g + SRh,$$

with moreover $\|h\|_2^2 = |c_0|^2 + \|Rh\|_2^2$, because the two functions on the right side in the preceding equality are orthogonal. Similarly, for any positive integer n, letting $c_n = \langle R^n h, g \rangle$, we have

$$R^n h = c_n g + SR^{n+1}h,$$

with $\|R^n h\|_2^2 = |c_n|^2 + \|R^{n+1}h\|_2^2$. Iterating, we obtain

$$h = (c_0 + c_1 S + \cdots + c_n S^n)g + S^{n+1}R^{n+1}h$$

for each positive integer n, with $\|h\|_2^2 = |c_0|^2 + |c_1|^2 + \cdots + |c_n|^2 + \|R^{n+1}h\|_2^2$. (The reasoning above paraphrases the first step in Hitt's original proof.)

It is now asserted that $\lim_{n\to\infty} R^n = 0$ in the strong operator topology. Granting that for the moment, we can pass to the limit in the preceding expression to obtain $h = gh'$, where h' is the function $\sum_0^\infty c_n z^n$, and $\|h'\|_2 = \|h\|_2$. Moreover, a calculation gives $Rh = gS^*h'$, showing that the subspace $M' = \{h/g : h \in M\}$ is S^*-invariant. This completes the proof of Hitt's theorem, except for the verification that $R^n \to 0$ strongly.

It is in that verification where de Branges–Rovnyak spaces enter. Let f be the outer part and v the inner part of the function g. Let the function b be the first member of the pair associated with f. By Proposition 5, the operator $T_{1-b}T_{\bar{f}}$ is an isometry of H^2 onto $\mathcal{H}(b)$. Hence the operator $T_{1-b}T_{\bar{g}}$ is a coisometry of H^2 onto $\mathcal{H}(b)$, its kernel being $\mathcal{H}(v)$. It turns out that the latter operator intertwines the operators R and $S^* : T_{1-b}T_{\bar{g}}R = S^*T_{1-b}T_{\bar{g}}$. From this intertwining and simple properties of $\mathcal{H}(b)$ one easily deduces the strong convergence of R^n to 0. Details, as noted earlier, are in [13].

If the S^*-invariant subspace M' in Hitt's theorem is the full space H^2, then g obviously must be an inner function, and M is an S-invariant subspace. If M' is proper then it equals $\mathcal{H}(u)$ for an inner function u. As g belongs to M, the space $\mathcal{H}(u)$ must contain the constant function 1, which means $u(0) = 0$. The question of how g and u must be related in order for T_g to act isometrically on $\mathcal{H}(u)$ now arises. The answer, which is given by the next proposition, was found in [13]. The proof is based on the equality in the corollary to Proposition 5.

Proposition 7. *Let g be a function of unit norm in H^2, and let b be the first member of the pair associated with the outer factor of g. Let u be an inner function with $u(0) = 0$. Then T_g acts isometrically on $\mathcal{H}(u)$ if and only if u divides b.*

(One should note that the function b in this proposition automatically vanishes at the origin, because of the condition $\|g\|_2 = 1$.)

It is worth examining briefly the special case of Hitt's theorem where $M = \mathcal{H}(v)$ with v an inner function satisfying $v(0) \neq 0$. The corresponding function g is then $(1 - |v(0)|^2)^{-1/2}(1 - \overline{v(0)}v)$, the kernel function in $\mathcal{H}(v)$ for evaluation at 0, divided by its norm. After some calculation one finds that the corresponding function b is given by

$$b = \frac{\overline{v(0)}(v(0) - v)}{1 - \overline{v(0)}v}.$$

With further calculations one can show that $g^{-1}\mathcal{H}(v) = \mathcal{H}(u)$ where $u = (v(0) - v)/(1 - \overline{v(0)}v)$. In this case, therefore, the inner function u corresponding to M is the inner part of b.

4. Hayashi's Theorem.

Let M be a nontrivial, proper, nearly S^*-invariant subspace of H^2, $M = T_g M'$, as in Hitt's theorem. We ask: When is M the kernel of a Toeplitz operator? Because kernels of Toeplitz operators are stable under division by inner functions, a necessary condition is that g be outer, in which case M will not be proper unless M' is. We may thus suppose, in keeping with notations used earlier, that $M = T_f \mathcal{H}(u)$, where f is an outer function of unit H^2 norm, u is an inner function that vanishes at the origin, and T_f acts isometrically on $\mathcal{H}(u)$. Let (b, a) be the pair associated with f. By Proposition 7, the inner function u divides b. We let $b_0 = b/u$ and $f_0 = \frac{a}{1-b_0}$. The pair (b_0, a) is then admissible. These notations will be retained throughout the present section.

Hayashi's Theorem. *The subspace M is the kernel of a Toeplitz operator if and only if the pair (b_0, a) is special and f_0^2 is rigid.*

The preceding statement differs superficially from Hayashi's but is easily seen to be equivalent. If (b_0, a) is special and f_0^2 is rigid then (b, a) is special and f^2 is rigid, by the corollary to Proposition 6. As mentioned in connection with the corollary, however, it is possible for f^2 to be rigid and f_0^2 not to be.

It is convenient to present the initial steps in the proof of Hayashi's theorem in the form of lemmas.

Lemma 1. *If M is the kernel of a Toeplitz operator then it is the kernel of $T_{\bar{u}\bar{f}/f}$.*

The proof from [13] will be repeated. Suppose φ is a function in L^∞ such that M is the kernel of T_φ. Because φ then multiplies a nonzero H^2 function into $(H^2)^\perp$, it is log-integrable and so can be written as $\varphi = \chi\bar{\psi}$, where ψ is the outer function with modulus $|\varphi|$ and χ is unimodular. Then $T_\varphi = T_{\bar{\psi}}T_\chi$, and since $T_{\bar{\psi}}$ has a trivial kernel, the kernel of T_φ equals that of T_χ. We may therefore assume that φ is unimodular.

Because $T_\varphi f = 0$, the function φf is in $(H^2)^\perp$, so (since it has the same modulus as f) it can be written as $\varphi f = \bar{u}_1 \bar{f}$ where u_1 is an inner function vanishing at the origin. Thus $\varphi = \bar{u}_1 \bar{f}/f$.

Suppose h is a function in $\mathcal{H}(u)$. Then fh is in $\ker T_\varphi$, implying that $\bar{u}_1 \bar{f}h$ is in $(H^2)^\perp$. Since f is an outer function it follows that $\bar{u}_1 h$ is in $(H^2)^\perp$, showing that $\bar{u}_1 \mathcal{H}(u) \subset (H^2)^\perp$. Thus u divides u_1.

Suppose h is a bounded function in $\mathcal{H}(u_1)$. Then

$$T_\varphi fh = P_+(\bar{u}_1 \bar{f}h) = 0,$$

implying that fh is in $T_f\mathcal{H}(u)$ and so that h is in $\mathcal{H}(u)$. Since $\mathcal{H}(u_1)$ is spanned by its bounded functions (for example, by its kernel functions), it follows that $\mathcal{H}(u_1) \subset \mathcal{H}(u)$, and so u_1 divides u.

Since u_1 is codivisible with u it is a constant multiple of u, and thus φ is a constant multiple of $\bar{u}\bar{f}/f$. This proves Lemma 1.

Lemma 2. *The operator $T_{1-b}T_{\bar{f}}$ acts on M as division by f : $T_{1-b}T_{\bar{f}}fh = h$ for h in $\mathcal{H}(u)$.*

Let h be in $\mathcal{H}(u)$. Then $T_{\bar{b}}h = 0$ because u divides b, so

$$T_{1-b}T_{\bar{f}}fh = T_{1-b}T_{\bar{f}}T_f T_{1-\bar{b}}h.$$

By the corollary to Proposition 5, the right side equals $(1 - T_b T_{\bar{b}})h$, which is h.

Lemma 3. *The operator $T_{1-b}T_{\bar{f}}$ maps the range of the operator $T_{uf/\bar{f}}$ onto uaH^2.*

This is a consequence of the equality

$$T_{1-b}T_{\bar{f}}T_{uf/\bar{f}} = T_{ua},$$

which follows by standard properties of Toeplitz operators.

Suppose now that M is the kernel of a Toeplitz operator, and thus by Lemma 1 the kernel of $T_{\bar{u}\bar{f}/f}$. By Lemma 3, the operator $T_{1-b}T_{\bar{f}}$ (an isometry of H^2 onto $\mathcal{H}(b)$) maps the range of the operator $T_{uf/\bar{f}}$ onto uaH^2, and by Lemma 2 it maps M (the orthogonal complement of the preceding range) onto $\mathcal{H}(u)$. By Proposition 4, the space $\mathcal{H}(b)$ is the orthogonal direct sum of $\mathcal{H}(u)$ and $u\mathcal{H}(b_0)$. Thus uaH^2 is dense in $u\mathcal{H}(b_0)$, relative to the norm of $\mathcal{H}(b)$. Also by Proposition 4, the operator T_u acts as an isometry of $\mathcal{H}(b_0)$ into $\mathcal{H}(b)$. Hence aH^2 must be dense in $\mathcal{H}(b_0)$, implying by Proposition 6 that (b_0, a) is special and f_0^2 is rigid. This establishes one half of Hayashi's theorem.

For the other direction, assume (b_0, a) is special and f_0^2 is rigid. Then, reversing the reasoning above, we can conclude that the orthogonal complement of the range of the operator $T_{uf/\bar{f}}$ is M, so that M is the kernel of $T_{\bar{u}\bar{f}/f}$. This completes the proof of the theorem.

The proof of Hayashi's theorem contains a recipe for constructing the general nontrivial proper subspace of H^2 that is the kernel of a Toeplitz operator. One starts with an outer function f_0 in H^2 such that f_0^2 is rigid, and an inner function u that vanishes at the origin. Let (b_0, a) be the pair associated with f_0. Let $b = ub_0$ and $f = \frac{a}{1-b}$. By the corollary to Proposition 6, the function f^2 is rigid, and the pair (b, a) is special. In particular, the square of the L^2 norm of f is the value of $\frac{1+b}{1-b}$ at the origin, which is 1. Thus $T_f \mathcal{H}(u)$ is one of Hitt's nearly S^*-invariant subspaces; by the proof of Hayashi's theorem it is the kernel of $T_{\bar{u}\bar{f}/f}$, and one obtains in this manner every nontrivial proper subspace of H^2 that is the kernel of a Toeplitz operator.

5. Rigid Functions.

Rigid functions in H^1 arise in several connections [1] and have received considerable attention in the past few years. However, we still lack a structural characterization of them. Recently, J. Inoue [9] has made progress by constructing a counterexample to a conjecture from [14].

Prior to Inoue's example, every known nonrigid outer function displayed a divisibility property involving inner functions. If v is a nonconstant inner function, then the functions v and $(1 + v)^2$ have the same argument almost everywhere on ∂D—this observation was used earlier, in Section 1. Hence, if G is an outer function in H^1 and if there is a nonconstant inner function v such that $G/(1+v)^2$ is in H^1, then G is not rigid, for it has the same argument almost everywhere on ∂D as $vG/(1+v)^2$. The conjecture made in [14] (actually, a part of the conjecture) was the converse of the preceding statement, namely, that if the outer H^1 function G is not rigid, then there is a nonconstant inner function v such that $G/(1+v)^2$ is in H^1. Although some positive evidence had accumulated [7], [13], [14], [15], Inoue's example refutes the conjecture. The example belongs to a new species of nonrigid function which we can hope will lead to a better understanding of rigid functions.

References.

1. P. Bloomfield, N. P. Jewell and E. Hayashi, *Characterizations of completely nondeterministic stochastic processes*, Pacific J. Math. **107** (1983), 307–317.
2. L. de Branges and J. Rovnyak, *Square Summable Power Series*, Holt, Rinehart and Winston, New York, 1966.
3. R. G. Douglas, *On majorization, factorization, and range inclusion of operators on Hilbert space*, Proc. Amer. Math. Soc. **17** (1966), 413–415.
4. E. Hayashi, *The solution sets of extremal problems in H^1*, Proc. Amer. Math. Soc. **93** (1985), 690–696.
5. E. Hayashi, *The kernel of a Toeplitz operator*, Integral Equations and Operator Theory **9** (1986), 588–591.
6. E. Hayashi, *On the classification of nearly invariant subspaces of the backward shift*, Proc. Amer. Math. Soc. **110** (1990), 441–448.
7. H. Helson, *Large analytic functions II*, Analysis and Partial Differential Equations, C. Sadosky (ed.), Marcel Dekker, New York (1990), 217–220.
8. D. Hitt, *Invariant subspaces of H^2 of an annulus*, Pacific J. Math. **134** (1988), 101–120.
9. J. Inoue, *An example of a non-exposed extreme function in the unit ball of H^1*, Proc. Edinburgh Math. Soc., forthcoming.
10. B. A. Lotto and D. Sarason, *Multiplicative structure of de Brange's spaces*, Rev. Mat.

Iberoamericana **7** (1991), 183–220.

11. D. Sarason, *Shift-invariant spaces from the Brangesian point of view*, The Bieberbach Conjecture – Proceedings of the Symposium on the Occasion of the Proof, Amer. Math. Soc., Providence (1986), 153–166.

12. D. Sarason, *Doubly shift-invariant spaces in H^2*, J. Operator Theory **16** (1986), 75–97.

13. D. Sarason, *Nearly invariant subspaces of the backward shift*, Operator Theory: Advances and Applications **35** (1988), 481–493.

14. D. Sarason, *Exposed points in H^1, I*, Operator Theory: Advances and Applications **41** (1989), 485–496.

15. D. Sarason, *Exposed points in H^1, II*, Operator Theory: Advances and Applications **48** (1990), 333–347.

Department of Mathematics
University of California
Berkeley, CA 94720

MSC 1991: 47B35

Operator Theory:
Advances and Applications, Vol. 71
© 1994 Birkhäuser Verlag Basel/Switzerland

A fixed point approach to Nehari's problem and its applications

SERGEI TREIL [1] AND ALEXANDER VOLBERG[2]

Dedicated to Harold Widom

We introduce a new approach to Nehari's problem. This approach is based on some kind of fixed point theorem and allows us to obtain some useful generalizations of Nehari's and Adamyan – Arov – Krein (AAK) theorems. Among those generalizations: descriptions of Hankel operators in weighted ℓ^2 spaces; descriptions of Hankel operators from Dirichlet type spaces to weighted Bergman spaces; commutant lifting theorem for non-contractive operators; AAK theorem for "four block operator" etc.

1. ABSTRACT NEHARI'S THEOREM

Let $\mathcal{H}_1$ and $\mathcal{H}_2$ be two separable Hilbert spaces, S_1 be an expanding operator ($\|S_1 x\| \geq \|x\| \; \forall x$) in $\mathcal{H}_1$ and S_2 be a contractive operator in $\mathcal{H}_2$. Let $\mathcal{H}_2 = \mathcal{H}_2^+ \oplus \mathcal{H}_2^-$ be a orthogonal decomposition of $\mathcal{H}_2$, and let

$$S_2 \mathcal{H}_2^+ \subset \mathcal{H}_2^+. \tag{1.1}$$

Denote by $\mathbb{P}_+$ and $\mathbb{P}_-$ orthogonal projections in $\mathcal{H}_2$ onto $\mathcal{H}_2^+$ and $\mathcal{H}_2^-$ respectively. A generalized Hankel operator is defined as a bounded linear operator Γ, $\Gamma : \mathcal{H}_1 \to \mathcal{H}_2^-$ satisfying the following commutant relations

$$\Gamma S_1 f = \mathbb{P}_- S_2 \Gamma f, \qquad f \in \mathcal{H}_1. \tag{1.2}$$

A bounded operator $T : \mathcal{H}_1 \to \mathcal{H}_2$ satisfying

$$T S_1 = S_2 T$$

· [1] Partially supported by NSF grant DMS 9304011
[2] Partially supported by NSF grant DMS 9101788

will be called (generalized) multiplier. Projections of multipliers give an example of Hankel operators; given a multiplier T one can construct Hankel operator Γ_T,

$$\Gamma_T f \stackrel{\text{def}}{=} \mathbb{P}_- T f, \qquad f \in \mathcal{H}_1. \tag{1.3}$$

It is easy to see that

$$\Gamma_T S_1 f = \mathbb{P}_- T S_1 f = \mathbb{P}_- S_2 T f = \mathbb{P}_- S_2 (\mathbb{P}_- + \mathbb{P}_+) T f = \mathbb{P}_- S_2 \mathbb{P}_- T f = \mathbb{P}_- S_2 \Gamma_T f$$

(here we use the fact that $\mathbb{P}_- S_2 \mathbb{P}_+ = 0$ provided by $S_2 \mathcal{H}_2^+ \subset \mathcal{H}_2^+$), so Γ_T is a Hankel operator. It is also clear that $\|\Gamma_T\| \leq \|T\|$.

Note, that in the above definitions we did not use the fact that the operator S_1 is expanding and S_2 is contractive. The following theorem that is the main result of this section shows us that under these assumption operators Γ_T are the only Hankel operators and the equality $\|\Gamma_T\| = \|T\|$ takes place (we also assume, of course, that (1.1) holds).

Theorem 1.1. *Let S_1 be an expanding operator ($\|S_1 x\| \geq \|x\| \ \forall x$) and S_2 be a contraction. Given Hankel operator Γ there exists a multiplier T (an operator $T : \mathcal{H}_1 \to \mathcal{H}_2$ satisfying $S_2 T = T S_1$) such that $\Gamma = \Gamma_T$ and moreover $\|\Gamma\| = \|T\|$.*

If S_1 and S_2 are shift operators (multiplications by independent variable z in H^2 and $L^2(\mathbb{T})$ respectively), then the theorem gives us classical Nehari's theorem [1] (Nehari – Page's theorem if S_1, S_2 are multiple shifts). The case of isometrical S_1 and S_2 gives us a generalized commutant lifting theorem considered in [2]. Other important examples and applications of the above abstract result will be presented in the next section.

The main new detail in our theorem is that we allow the operator S_1 to be an expanding operator, not necessarily an isometry.

Let us discuss the methods of proving Nehari type theorems. There are known several types of such methods. The first type is based on original Nehari's proof [1] using a duality. In scalar case it uses the factorization result that the products $f_1 f_2$ where f_1, f_2 are functions in the unit ball of H^2 form a dense subset of the unit ball of H^1. It also uses the fact that $(H^1)^* = L^\infty / H^1_0$. This approach can be used also in vector case.

The second approach is based on the so called one step extension by Parrot theorem, see [3]. This approach was used by Adamyan – Arov – Krein [5]. This approach unfortunately does not work in our situation, because it uses the fact that the shift operator S, $Sf(z) = zf(z)$ is an isometry.

A very interesting approach based on the lifting of the so-called Generalized Toeplitz Kernels was treated by R. Arocena, M. Cotlar and C Sadosky [6]. Nehari's theorem is simply a partial case of their general result about such kernels. But their approach uses either some kind of factorization property mentioned above or the fact that the shift operator S is an isometry, cf [6, 7, 8]. Such an approach allows one to obtain Nehari type theorems for Hankel operators in weighted L^2 spaces, or for Hankel operators on tori see [9], but unfortunately it again can not be used in our situation.

Another approach was introduced by Ball and Helton [10]. Their approach is essentially geometrical and uses the setting of Krein spaces (spaces with indefinite inner product). They reduce the Nehari's theorem to the problem of finding a maximal nonnegative

z-invariant subspace in (a subspace of) some Krein space. To prove the existence of such a subspace they used a Beurling – Lax type description of all z-invariant subspaces M of $L^2(E)$. The classical Beurling – Lax theorem says that any z-invariant, simply invariant (i.e. such that $\cap z^n M = \{0\}$) subspace M of $L^2(E)$ can be represented as $M = gH^2(E')$ where the operator function g, $g(z) : E' \to E$, $z \in \mathbb{D}$ is an isometry a.e. on $\mathbb{T}$.

The main idea in the approach of Ball and Helton is that the representation $M = gH^2(E')$ is highly non unique in general. The assumption that g is an isometry makes it unique up to a constant operator multiplier. Ball and Helton used their own nonclassical Beurling – Lax type theorem that claims that under some additional condition there is a representation with a J-isometrical g, where J gives the Krein structure on $L^2(E)$. Theorems of such type are unknown in our situation.

Our approach below does not use a description of the invariant subspaces. To prove the existence of the maximal nonnegative invariant subspace we use a version of fixed point theorem by Ky Fan [21] and I. S. Iokhvidov [20]. This theorem was used by one of the authors [11] for the proof of the vector version of the Adamyan – Arov – Krein theorem. The main step of this proof, where the above fixed point theorem was used was the reduction of the problem to the Nehari's theorem. Recently the second author noticed that this fixed point theorem can be used for the proof of Nehari's theorem (and its generalizations) itself.

There is nothing specifically from the "Hilbert space theory" in the application of this fixed point theorem, although the reduction to it uses indefinite inner product spaces and Hilbert structure. This is what prevents us from easy extension of our technique on the Hankel operators in ℓ^p, $p \neq 2$.

In conclusion of this introduction we would like to thank the referee for several useful remarks.

2. MAIN EXAMPLES AND COROLLARIES

The first three examples presented in this section (Sections 2.1 — 2.3) also work in the vector-valued case, when we consider spaces of functions with values in separable Hilbert spaces. For the simplicity of notation we present only the scalar-valued case, but all reasonings are absolutely the same in the vector case.

2.1. OPERATORS FROM DIRICHLET TYPE SPACES TO WEIGHTED BERGMAN SPACES

In this example $\mathcal{H}_1$ be a space of analytic functions $f = \sum_{n \geq 0} \hat{f}(n) z^n$ on the unit disk $\mathbb{D}$ such that

$$\|f\|_{\mathcal{H}_1}^2 \overset{\text{def}}{=} \sum_{n \geq 0} |\hat{f}(n)|^2 u_n < \infty.$$

Weights $\{u_n\}$ is supposed to be increasing, $u_n \leq u_{n+1} \leq \dots$ and satisfying

$$\sup_{n \geq 0} \frac{u_{n+1}}{u_n} < \infty.$$

The operator S_1 will be multiplication by z. Its boundedness is guaranteed by the above inequality, and it is expanding because weight $\{u_n\}$ is increasing.

The space $\mathcal{H}_2$ will be the weighted space $L^2(\mu)$ where μ is a finite Borel measure on the closed unit disk clos$\mathbb{D}$, and S_2 again will be multiplication by z here. Denote by $A^2[\mu]$ the weighted Bergman space, i.e. the closure of analytic polynomials in $L^2(\mu)$-norm,

$$A^2[\mu] \stackrel{\text{def}}{=} \text{clos}_{L^2(\mu)} \text{span} \, (z^n : n \geq 0).$$

We will consider two types of Hankel operators ("big" and "small") that correspond to the following two decompositions of $\mathcal{H}_2$.

"Big" Hankel operators: $\mathcal{H}_2^+ = A^2[\mu]$ and $\mathcal{H}_2^- = L^2(\mu) \ominus A^2[\mu]$.

"Small" Hankel operators: $\mathcal{H}_2^-$ is the complex conjugate to the Bergman space $A^2[\mu]$, $\mathcal{H}_2^- = \overline{A^2[\mu]}$, and $\mathcal{H}_2^+ = L^2(\mu) \ominus \overline{A^2[\mu]}$ (bar means complex conjugation here).

The Bergman space $A^2[\mu]$ is clearly z-invariant $(S_2$-invariant$)$. To show that for "small" Hankel operators the subspace $\mathcal{H}_+ = L^2(\mu) \ominus \overline{A^2[\mu]}$ is S_2-invariant let us note that the adjoint operator S_2^* is the multiplication by $\bar{z}$ in $L^2(\mu)$ and the space $\mathcal{H}_2^- = \overline{A^2[\mu]}$ is $\bar{z}$-invariant.

Multipliers in this case are simply multiplications by some $L^2(\mu)$ functions. Note, that an arbitrary function $f \in L^2(\mu)$ does not necessarily generate a (bounded) multiplier, but any multiplier belongs to $L^2(\mu)$.

Given a generalized Hankel operator Γ (an operator satisfying (1.2)) one can define a function $\varphi \in L^2(\mu)$ — an "antianalytic part of symbol" by $\varphi \stackrel{\text{def}}{=} \Gamma 1$. Then the action of Γ on polynomials can be rewritten as

$$\Gamma f = \mathbb{P}_- \varphi f, \qquad f \in \mathcal{H}_1 .$$

Any symbol of the Hankel operator Γ has the same "antianalytic part", i.e. if $\Gamma = \Gamma_\psi$ then $\mathbb{P}_- \psi = \varphi$.

Theorem 1.1 implies that a Hankel Γ is bounded if and only if there exists a symbol which is a multiplier from $\mathcal{H}_1$ to $L^2(\mu)$ and, moreover, the norm of Γ equals the minimal possible norm of such a multiplier (i.e. the minimum is attained). Note, that this is true for both "big" and "small" Hankel operators.

The description of multipliers is especially simple if $u_n \equiv 1$. In that case $\mathcal{H}_1 = H^2$ and a function φ is a multiplier if and only if the measure $|\varphi|^2 d\mu$ is Carleson. In the case of nontrivial weights the description of multipliers also can be given in terms of corresponding imbedding theorems, but we do not want to discuss such imbedding theorems here.

2.2. CLASSICAL HANKEL OPERATORS IN WEIGHTED ℓ^2 SPACES

Let $\{u_n\}_{n\geq 0}$ be a sequence of positive numbers. Consider a weighted space $\ell^2(\{u_n\})$

$$\ell^2(\{u_n\}) \stackrel{\text{def}}{=} \left\{ \{x_n\}_{n=0}^{\infty} : \sum_{n=0}^{\infty} |x_n|^2 u_n < \infty \right\},$$

and let $\ell^2(\{v_n\})$ be another such space. We will call a *bounded* operator Γ acting from $\ell^2(\{u_n\})$ to $\ell^2(\{v_n\})$ Hankel (classical Hankel operator) if its matrix $(\gamma_{j,k})_{j,k=0}^{\infty}$ in the standard bases in ℓ^2 has Hankel structure, $\gamma_{j,k} = \gamma_{j+k}$. In other words,

$$\Gamma\{x_j\}_{j=0}^{\infty} = \left\{ \sum_{j=0}^{\infty} \gamma_{j+k} x_j \right\}_{k=0}^{\infty}. \tag{2.4}$$

We will consider the case when both weights $\{u_n\}$ and $\{v_n\}$ are increasing, $u_n \leq u_{n+1}$, $v_n \leq v_{n+1}$. We also suppose that as in Section 2.1

$$\sup_{n\geq 0} \frac{u_{n+1}}{u_n} < \infty.$$

In this case Theorem 1.1 gives us a description of all Hankel operators. Let us define w_n, $n \in \mathbb{Z}$ by the formula

$$w_n = \begin{cases} v_{-n}, & n \leq 0 \\ \\ v_0, & n \geq 0 \end{cases}$$

Consider the weighted space $\mathcal{H}_2 = \ell^2(\{w_n\})$ of all all bilateral sequences $\{x_n\}_{n=-\infty}^{\infty}$ satisfying $\sum_{n=-\infty}^{\infty} |x_n|^2 w_n < \infty$ and let

$$\mathcal{H}_2^+ \stackrel{\text{def}}{=} \left\{ \{x_n\} \in \ell^2(\{w_n\}) : x_n = 0 \text{ for } n \leq 0 \right\},$$
$$\mathcal{H}_2^- \stackrel{\text{def}}{=} \left\{ \{x_n\} \in \ell^2(\{w_n\}) : x_n = 0 \text{ for } n > 0 \right\}.$$

A sequence $a = \{a_k\}_{k=-\infty}^{\infty}$ is called a multiplier from $\ell^2(\{u_n\})$ to $\ell^2(\{w_n\})$ if the operator L_a,

$$L_a\{x_j\}_{j=0}^{\infty} = \left\{ \sum_{j=0}^{\infty} a_{k-j} x_j \right\}_{k=-\infty}^{\infty}.$$

(which is also called a multiplier) is bounded. In the case of trivial weights ($u_n \equiv 1$, $w_n \equiv 1$) a is a multiplier if and only if $\sum_{j=-\infty}^{\infty} a_j z^j$ is bounded on the unit circle $\mathbb{T}$, and the norm of corresponding operator (the norm of multiplier) coincides with the L^{∞}-norm.

Consider a multiplier L_a. Its compression $\Gamma_L = \mathbb{P}_- L_a : \ell^2(\{u_n\}) \to \mathcal{H}_2^-$ (recall that $\mathbb{P}_- = P_{\mathcal{H}_2^-}$) has in standard bases in ℓ^2 a Hankel structure (2.4) with $\gamma_k = a_{-k}$.

Let S_1 and S_2 be shift operators in $\ell^2(\{u_n\})$ and $\ell^2(\{w_n\})$ respectively, $S_{1,2}e_n = e_{n+1}$, e_n be standard basis elements in ℓ^2. Let Γ be a bounded Hankel operator, $\Gamma : \ell^2(\{u_n\}) \to \mathcal{H}_2^-$; we mean that its matrix in standard bases in ℓ^2 has Hankel structure (2.4). This means exactly that $\Gamma S_1 = \mathbb{P}_- S_2 \Gamma$. So operators Γ, S_1, S_2 satisfy the hypotheses of Theorem 1.1, and therefore Γ admits a lifting to a multiplier L. In other words there exists a multiplier L, $\|L\| = \|\Gamma\|$ such that $\Gamma = \Gamma_L$.

This result seems to be not so interesting, since the space of multipliers of weighted ℓ^2-spaces is a mysterious object. But if we impose some additional assumption on the weights, we can obtain a more reasonable description of Hankel operators.

2.3. CLASSICAL HANKEL OPERATORS IN WEIGHTED ℓ^2 SPACES, II

Let us now consider weighted spaces $\ell^2(\{u_n\})$ and $\ell^2(\{v_n\})$ with additional assumptions about the weights. Namely, as in Section 2.1 we assume about the weight u_n that it is increasing and

$$\sup_{n \geq 0} \frac{u_{n+1}}{u_n} < \infty.$$

As to v_n, we assume that

$$\frac{1}{v_n} = \int_0^1 x^{2n} d\nu(x),$$

where ν is some finite nonnegative Borel measure on $[0, 1]$.

Let Γ be a bounded Hankel operator acting from $\ell^2(\{u_n\})$ to $\ell^2(\{v_n\})$,

$$\Gamma\{x_j\}_{j=0}^\infty = \left\{ \sum_{j=0}^\infty \gamma_{j+k} x_j \right\}_{k=0}^\infty.$$

Let us denote by μ a *radial symmetrization of* ν, that is a radially symmetric measure μ on $\operatorname{clos} \mathbb{D}$ such that

$$\int_{\operatorname{clos} \mathbb{D}} |z|^{2n} d\mu(z) = \int_0^1 x^{2n} d\nu(x). \tag{2.5}$$

Let us define φ as

$$\varphi = \sum_0^\infty \gamma_n v_n \bar{z}^n$$

Using the mapping $\{a_n\}_0^\infty \mapsto \sum_0^\infty a_n z^n$ one can identify $\ell^2(\{u_n\})$ with the space $\mathcal{H}_1$ of analytic functions in the unit disk $\mathbb{D}$ described in Section 2.1, and using the mapping $\{a_n\}_0^\infty \mapsto \sum_0^\infty a_n \bar{z}^n$ we identify $\ell^2(\{1/v_n\})$ with the antianalytic weighted Bergman space $A^2[\mu]$ (closure of antianalytic polynomials in $L^2(\mu)$, see Section 2.1). (Due to the radial symmetry of μ the functions $\bar{z}^n$ are orthogonal, so (2.5) implies that the last mapping is an isometry.)

Boundedness of Γ implies $\{\gamma_n\}_0^\infty \in \ell^2(\{v_n\})$, so $\{\gamma_n v_n\}_0^\infty \in \ell^2(\{1/v_n\})$, or, equivalently $\varphi \in L^2(\mu)$. Now

$$\|\Gamma\| = \sup \left\{ \left| \sum_{0 \leq j,k \leq 0} \gamma_{j+k} x_j \bar{y}_k \right| : \begin{array}{l} \{x_j\} \in \ell^2(\{u_n\}),\ \|\{x_j\}\| \leq 1, \\ \{y_j\} \in \ell^2(\{1/v_n\}),\ \|\{y_j\}\| \leq 1 \end{array} \right\} =$$

$$= \sup \left\{ \left| \int_{\mathrm{clos}\,\mathbb{D}} \varphi f \bar{g} d\mu \right| \ : \ \begin{array}{c} f \in \mathcal{H}_1, \ \|f\| \leq 1, \\ g \in \overline{A^2[\mu]}, \ \|g\| \leq 1 \end{array} \right\} = \sup_{f \in \mathcal{H}_1, \|f\| \leq 1} \|\Gamma_\varphi f\| = \|\Gamma_\varphi\|,$$

where Γ_φ is a "small" Hankel operator acting from Dirichlet type space $\mathcal{H}_1$ to antianalytic weighted Bergman space $\overline{A^2[\mu]}$ described in Section 2.1, $\Gamma_\varphi f = P_{\overline{A^2[\mu]}} \varphi f$.

So we can apply the result of Section 2.1 to obtain that Hankel operator Γ_φ admits a lifting to a multiplier from $\mathcal{H}_1$ to $L^2(\mu)$. In other words the operator Γ is bounded if and only if there exists a function $\psi \in L^2(\mu)$ such that

$$\int_{\mathrm{clos}\,\mathbb{D}} z^n \varphi d\mu = \int_{\mathrm{clos}\,\mathbb{D}} z^n \psi d\mu = \gamma_n, \qquad n \geq 0,$$

and the measure $|\psi|^2 d\mu$ is a Carleson measure for $\mathcal{H}_1$, i.e. the imbedding

$$\int_{\mathrm{clos}\,\mathbb{D}} |f|^2 |\psi|^2 d\mu \leq C \|f\|_{\mathcal{H}_1}^2 \qquad \forall f \in \mathcal{H}_1$$

holds. Moreover, the norm of Γ equals the least possible (over all ψ) norm of the imbedding operator.

This result is especially simple when $u_n \equiv 1$, so $\mathcal{H}_1$ is usual Hardy space H^2, and measure $|\psi|^2 d\mu$ has to be simply a classical Carleson measure.

2.4. A NEW COMMUTANT LIFTING THEOREM

Let us consider two operators T_1 and T_2 in Hilbert spaces H_1 and H_2 respectively, and let T_2 be a contraction. Let $S_1 : \mathcal{H}_1 \to \mathcal{H}_1$ be an expanding ($\|Sx\| \geq \|x\| \ \forall x$) dilation of T_1 and $S_2 : \mathcal{H}_2 \to \mathcal{H}_2$ be a contractive dilation of T_2.

Here by dilation we mean that $H_{1,2} \subset \mathcal{H}_{1,2}$ and

$$S_{1,2}^* H_{1,2} \subset H_{1,2}, \qquad T_{1,2} = P_{H_{1,2}} S_{1,2} \mid H_{1,2}.$$

Note, that the above equalities imply

$$T_{1,2}^n = P_{H_{1,2}} S_{1,2}^n \mid H_{1,2}, \qquad n \geq 0,$$

so the definition is similar to the classical one. The only difference is that for classical dilations inclusions $S_{1,2}^* H_{1,2} \subset H_{1,2}$ do not necessarily hold. But if we consider, as in the classical commutant lifting theorem, minimal isometric dilations, then the above inclusions take place. So the following result can be considered as a generalization of the commutant lifting theorem

Theorem 2.1. *Let T_1 and T_2 be operators in H_1 and H_2 and let S_1 and S_2 be (respectively) their expanding and contractive dilations described above. Let $A : H_1 \to H_2$ be a bounded operator such that*

$$AT_1 = T_2 A.$$

Then there exists a bounded operator $T : \mathcal{H}_1 \to \mathcal{H}_2$, $TS_1 = S_2T$, such that

$$A = P_{H_2}T \mid H_1, \qquad TH_1^\perp \subset H_2^\perp$$

and, moreover $\|T\| = \|A\|$.

If S_1 and S_2 are minimal isometric dilations of T_1 and T_2 the above theorem gives us a classical commutant lifting theorem [19]. But our theorem can be also applied to non-contractive operators T_1 and their expanding dilations.

Proof of Theorem 2.1. Let us define $\Gamma : \mathcal{H}_1 \to H_2$ to be A on H_1 and 0 on $\mathcal{H}_1 \ominus H_1$. Taking into account that $T_{1,2} = P_{H_{1,2}}S_{1,2} \mid H_{1,2}$ and $S_{1,2}H_{1,2}^\perp \subset H_{1,2}^\perp$ one can rewrite the equality $AT_1 = T_2A$ as

$$\Gamma S_1 = P_{H_2}S_2\Gamma.$$

So Γ is a generalized Hankel operator and satisfies all assumptions of Theorem 1.1 (with $\mathcal{H}_2^- = H_2$).

Therefore there exists an operator $T : \mathcal{H}_1 \to \mathcal{H}_2$, $TS_1 = S_2T$ such that $\Gamma = P_{H_2}T$ (so $A = P_{H_2}T \mid H_2$) and, moreover $\|A\| = \|\Gamma\| = \|T\|$. Note that $\Gamma = P_{H_2}T$ implies $TH_1^\perp \subset H_2^\perp$. ●

3. IOKHVIDOV – KY FAN THEOREM AND SOME ITS APPLICATIONS: A GENERALIZED AAK THEOREM

The main technical tool used to prove the main result (Theorem 1.1) is the following theorem by I. S. Iokhvidov – Ky Fan.

Let A be a selfadjoint operator in a separable Hilbert space $\mathcal{H}$, $\mathcal{P}_+$ and $\mathcal{P}_-$ be orthogonal projection onto nonnegative and (strictly) negative parts of its spectrum. We will assume in what follows that the operator

$$A_- = \mathcal{P}_- A \mid \mathcal{P}_- \mathcal{H} \tag{3.1}$$

is invertible. Let

$$\mathcal{K}_+ = \mathcal{K}_+^A \overset{\text{def}}{=} \{ f \in \mathcal{H} : (Af, f) \geq 0 \}$$

be the cone of A-nonnegative vectors.

Theorem 3.1. (I. S. Iokhvidov – Ky Fan [20]) *Let A_- be invertible and let S be a bounded operator in $\mathcal{H}$ such that*

$$S\mathcal{K}_+ \subset \mathcal{K}_+ \tag{3.2}$$

$$\mathcal{P}_+ S \mathcal{P}_- \quad \text{is a compact operator} \tag{3.3}$$

Then there exists a subspace M of $\mathcal{K}_+$ which is maximal (by inclusion) and S-invariant: $SM \subset M$.

3.1. GENERALIZED AAK THEOREM

The above Theorem 3.1 allows us also to obtain the following version of Adamyan – Arov – Krein (AAK) Theorem [4] for generalized Hankel operators defined in Section 1.

Let us remind the reader that for an operator A in a Hilbert space the singular number (s-number) $s_n = s_n(A)$ is defined as

$$s_n(A) = \inf\{\|A - K\| : \text{rank } K \leq n\}.$$

Another equivalent definition (for operators in a Hilbert space)

$$s_n(A) = \inf\{\|A \mid M\| : \text{codim } M \leq n\} = \tag{3.4}$$

$$= \sup_{L: \dim L \leq n+1} \inf_{x \in L, \|x\|=1} \|Ax\| \tag{3.5}$$

Theorem 3.2. *Let Γ be a generalized Hankel operators defined in Section 1, and let $s_n = s_n(\Gamma)$, $n \geq 0$ be the sequence of its s-numbers. Then*

$$s_n(\Gamma) = \inf\left\{\|\Gamma \mid M\| : M, \text{ codim } M \leq n, \ S_1 M \subset M\right\}$$

and the infimum *is attained.*

Note, that the inequality "$\leq$" above is a trivial consequences of (3.4).

As one can easily see the restriction of a generalized Hankel operator onto an S_1-invariant subspace is also a generalized Hankel operator. Therefore we can apply Theorem 1.1 to the operator $\Gamma \mid M$ from the above theorem, to obtain its lifting to a multiplier from M to $\mathcal{H}_2$. So Theorem 3.2 can be reformulated as follows:

Theorem 3.3. *Let again Γ be a generalized Hankel operators as defined in Section 1, and let $s_n = s_n(\Gamma)$, $n \geq 0$ be the sequence of its s-numbers. Then*

$$s_n(\Gamma) = \inf_{\substack{M, \text{ codim } M \leq n, \\ S_1 M \subset M}} \inf\{\|L\| : L \text{ is multiplier from } M \text{ to } \mathcal{H}_2, \ \mathbb{P}_- L = \Gamma \mid M \}$$

and the infimum *is attained.*

Proof of Theorem 3.2. To prove the theorem let us apply Theorem 3.1 to the operator $A = s_n^2 I - \Gamma^*\Gamma$, the space $\mathcal{H} = \mathcal{H}_1$ (s_n is nth singular number of the operator Γ) and the operator $S = S_1$. The operator A_- is clearly invertible because rank $\mathcal{P}_- \leq n < \infty$. The operator $\mathcal{P}_+ S \mathcal{P}_-$ has rank at most n so it is compact. Let us consider the cone $\mathcal{K}_+ = \mathcal{K}_+^A$ of A-nonnegative vectors,

$$\mathcal{K}_+ = \{f : (Af, f) \geq 0\} = \{f : \|\Gamma f\| \leq s_n \|f\|\}.$$

Since $\|S_1 f\| \geq \|f\|$ and $\|S_2 g\| \leq \|g\|$ for all $f \in \mathcal{H}_1$, $g \in \mathcal{H}_2$, we have for any $f \in \mathcal{K}_+$

$$\|\Gamma S_1 f\| = \|P_- S_2 \Gamma f\| \leq \|S_2 \Gamma f\| \leq \|\Gamma f\| \leq s_n \|f\| \leq s_n \|S_1 f\|.$$

Therefore $S_1 \mathcal{K}_+ \subset \mathcal{K}_+$ and all assumptions of Theorem 3.1 are satisfied. Hence there exists a maximal non-negative subspace $M \subset \mathcal{K}_+^A$ such that $S_1 M \subset M$. To complete the proof it remains to notice that $\operatorname{codim} M = \operatorname{rank} A_- \le n$ for any maximal subspace $M \subset \mathcal{K}_+^A$, cf. below Lemma 5.1, statement 4. ●

3.2. SINGULAR NUMBERS OF A "FOUR BLOCK OPERATOR"

In this section $L^2(E)$ denotes usual L^2 space on the unit circle $\mathbb{T}$ with values in a separable Hilbert space (all Hilbert spaces are supposed to be separable), $H^2(E)$ denotes vector Hardy space of analytic functions, $H_-^2(E)$ — its orthogonal complement, $H_-^2(E) = L^2(E) \ominus H^2(E)$. Symbols P_+, P_- are used for orthogonal projections onto $H^2(E)$ and $H_-^2(E)$ respectively.

Let us consider the following so-called "four block operator" $K : H^2(E_1) \oplus L^2(E_2) \to H_-^2(E_3) \oplus L^2(E_4)$ (all E_k are separable Hilbert spaces)

$$K = \begin{pmatrix} \Gamma_\Phi & P_- R \\ Q & F \end{pmatrix} \tag{3.6}$$

Here Γ_Φ is a usual vectorial Hankel operator $\Gamma_\Phi : H^2(E_1) \to H_-^2(E_3)$,

$$\Gamma_\Phi f = P_- \Phi f, \qquad f \in H^2(E_1)$$

(without loss of generality one can assume that Φ is a bounded function), and R, Q, F are multiplications by bounded operator-valued functions.

This operator plays an important role in control theory (H^∞ optimization). It arises from the so-called "model matching problem" to which the standard H^∞ optimization problem can be reduced, see [17].

If we define spaces $\mathcal{H}_1$, $\mathcal{H}_2$, $\mathcal{H}_2^-$ as

$$\mathcal{H}_1 = H^2(E_1) \oplus L^2(E_2)$$

$$\mathcal{H}_2 = L^2(E_3) \oplus L^2(E_4), \qquad \mathcal{H}_2^- = H_-^2(E_3) \oplus L^2(E_4)$$

and operators S_1 and S_2 are multiplications by the independent variable z in direct sums $\mathcal{H}_1$ and $\mathcal{H}_2$ respectively, then it is easy to see that the "four block operator" is a generalized Hankel operator. So we can apply Theorem 1.1. Taking into account that

$$\mathbb{P}_- = P_{\mathcal{H}_2^-} = \begin{pmatrix} P_- & 0 \\ 0 & I \end{pmatrix}$$

we obtain that

$$\|K\| = \inf \left\{ \left\| \begin{pmatrix} \Psi & R \\ Q & F \end{pmatrix} \right\| : \Psi, \Phi - \Psi \in H^\infty \right\} \tag{3.7}$$

and the infimum is attained. The above result (3.7) is well known and can be obtained by different methods. For example one step extension works in this case.

We are going to present a generalization of AAK theorem for "four block operators".

Let $H^\infty + \mathcal{R}_n = H^\infty + \mathcal{R}_n(E \to E_*)$ (usually we will omit $E \to E_*$) be the set of all operator-valued functions $F \in L^\infty(E \to E_*)$ (whose values are bounded operators from E to E_*) admitting representation $F = H + R$, where $H \in H^\infty(E \to E_*)$ and R is a rational matrix-valued function with poles in $\mathbb{D}$ and of MacMillan degree at most n.

For scalar-valued functions the MacMillan degree is exactly the number of poles (counting multiplicities). In matrix-valued case the definition is the same if we define the multiplicity of a pole for matrix-valued function in a right way. Without going into further details we will present two equivalent definitions of $H^\infty + \mathcal{R}_n$.

Class $H^\infty + \mathcal{R}_n$ consists of all functions F admitting the representation $F = HB^{-1}$, where $H \in H^\infty(E \to E_*)$ and $B \in H^\infty(E \to E)$ is a Blaschke – Potapov product of degree at most n, $\deg B \stackrel{\text{def}}{=} \dim H^2 \ominus BH^2$.

Another equivalent definition: class $H^\infty + \mathcal{R}_n$ is exactly the set of all $F \in L^\infty(\mathbb{T})$ such that $\operatorname{rank} H_F \le n$. For more information about these definitions and a constructive formula for calculation of the multiplicity of a pole of matrix valued functions see [11].

Theorem 3.4. *Let K be four block operator* (3.6), *and $s_n = s_n(K)$ be the sequence of its singular numbers. Then*

$$s_n(K) = \inf\left\{ \left\| \begin{pmatrix} \Psi & R \\ Q & F \end{pmatrix} \right\| : \Psi,\ \Phi - \Psi \in H^\infty + \mathcal{R}_n \right\}$$

and the infimum is attained.

If we have only one nontrivial block Γ_Φ the above theorem is the AAK theorem for matrix-valued functions, see [10, 11]. If $n = 0$ then the theorem gives us (3.7).

The particular case of this theorem when all functions Φ, R, Q, F are scalar-valued was proved in [12]. The proof there was similar to the original proof of Adamyan – Arov – Krein. The general result was proved (using methods different from ours) in [13, 14]. We would like to thank the referee who informed us about these papers.

Proof of Theorem 3.4. The upper estimate for s_n is trivial. To prove the opposite estimate let us apply to K Theorem 3.3. We obtain a S_1-invariant subspace M, $\operatorname{codim} M \le n$ and a multiplier $L : M \to \mathcal{H}_2 = L^2(E_3) \oplus L^2(E_4)$ such that

$$K\,\big|\,M = \mathbb{P}_- L \qquad \text{and} \qquad \|L\| = s_n(K).$$

Any invariant subspace of $\mathcal{H}_1 = H^2(E_1) \oplus L^2(E_2)$ of codimension at most n has a form

$$M = BH^2(E_1) \oplus L^2(E_2),$$

where B is a finite Blaschke – Potapov product of degree at most n. For us the essential property of finite Blaschke – Potapov products is that their values on the unit circle $\mathbb{T}$ are unitary operators.

The equality $K \mid M = \mathbb{P}_- L$ implies that the multiplier L has a form

$$L = \begin{pmatrix} \Psi & R \\ Q & F \end{pmatrix}$$

where Ψ satisfies

$$\Gamma_\Phi \mid BH^2(E_1) = \Gamma_\Psi \mid BH^2(E_1)$$

or, equivalently

$$P_- \Phi B f = P_- \Psi B f, \qquad \forall f \in H^2(E_1).$$

It can be rewritten as

$$\Phi B - \Psi B \in H^\infty,$$

or, what is equivalent, $\Phi - \Psi \in H^\infty + \mathcal{R}_n$ (here we use the fact that B is a unitary valued function). $\bullet$

4. Proof of the generalized Nehari's theorem

4.1. Hilbert spaces with indefinite inner product

The proof of the main theorem uses some simple ideas from the theory of Krein spaces (more precisely, the theory of indefinite inner product spaces). Although it is possible to write the proof without even mentioning Krein spaces and indefinite metric, we do not want to hide the ideas.

We do not assume that the reader knows the theory of indefinite inner product spaces or Krein spaces. All necessary information for understanding the proof he will find below. The following sources could serve as references for the theory of such spaces: [22, 23].

We call the attention of the experts to the fact that we do not even consider general indefinite inner product spaces. We deal only with the situation when an underlying Hilbert space is prespecified and indefinite inner product is introduced with the help of a selfadjoint operator in this space.

Namely, let $\mathcal{H}$ be a separable Hilbert space and J be a selfadjoint operator on $\mathcal{H}$. We can define on $\mathcal{H}$ an indefinite inner product $[.,.]$ generated by J as

$$[x, y] = (Jx, y), \qquad x, y \in \mathcal{H}.$$

We shall call the space $\mathcal{H}$ endowed with an indefinite inner product $[.,.]$ an *indefinite Hilbert space* to emphasize the fact that besides indefinite inner product it also has a regular Hilbert space structure. If J is invertible then this indefinite Hilbert space is called a *Krein space*.

Usually, see [23, 22], indefinite inner product space is defined axiomatically and it is not supposed apriori that it also has a regular Hilbert space metric. Moreover it is not always possible to represent an indefinite inner product space (defined axiomatically) as was described above. A special term "W-spaces" is used in the literature, see [22] for what we are calling here "indefinite Hilbert spaces". We don't like the former term and we think the later is much more informative. [3]

Let $\mathcal{H}_+$ and $\mathcal{H}_-$ be the spectral subspaces corresponding to the nonnegative and (strictly) negative parts of spectrum of J. Note that $\mathrm{Ker}\, J \subset \mathcal{H}_+$. Denote by $\mathcal{P}_\pm$ the orthogonal projections onto $\mathcal{H}_\pm$, and let $J_\pm = J|\mathcal{H}_\pm$.

Using the orthogonal decomposition

$$x = x_+ \oplus x_-, \qquad x_\pm \in \mathcal{H}_\pm,$$

one can rewrite the indefinite inner product $[x, y] = (Jx, y)$ as

$$[x, y] = [x_+, y_+] + [x_-, y_-] = (Jx_+, x_+) + (Jx_-, x_-) = (J_+ x_+, x_+) + (J_- x_-, x_-).$$

We will call the decomposition $\mathcal{H} = \mathcal{H}_+ \oplus \mathcal{H}_-$ *canonical*. The above formula is a characteristic property of the canonical decomposition, see Lemma 4.1 below.

As usual a vector $x \in \mathcal{H}$ is called positive (resp. negative, nonpositive, nonnegative) if $[x, y] > 0$ (resp $<, \leq, \geq$).

A subspace is called nonnegative if it consists of nonnegative vectors and is called negative if all vectors in it except 0 are negative. A nonnegative subspace M is called maximal nonnegative if the only nonnegative subspace containing M is M itself.

Lemma 4.1. *Let $\mathcal{H} = H_+ \oplus H_-$ be an orthogonal decomposition of an indefinite inner product space $\mathcal{H}$. Let the subspace H_+ be nonnegative, H_- be (strictly) negative, and let*

$$(J(x_+ + x_-), x_+ + x_-) = (Jx_+, x_+) + (Jx_-, x_-) \qquad \forall x_+ \in H_+, \forall x_- \in H_-. \qquad (4.1)$$

Then the decomposition $\mathcal{H} = H_+ \oplus H_-$ is canonical, i.e. the spaces $H_\pm$ are the spectral spaces of the operator J.

Remark. We would like to call the attention of the experts that our canonical decomposition depends on both indefinite inner product $[.,.]$ and the Hilbert space inner product $(.,.)$. It is unique if both inner products are fixed. So it is quite different from the "fundamental decomposition" for a Krein space as defined for example in [23] (which is non-unique).

Proof of Lemma 4.1. The equality (4.1) implies that

$$(Jx_+, x_-) = 0 \qquad \forall x_+ \in H_+, \forall x_- \in H_-.$$

---

[3]The term "W-spaces" is not as widespread as many similar, K-theory for example. So we have a courage to follow the referee's advise and to introduce our own term. We did not intend to do so because really we don't need indefinite inner product spaces and the only reason why we introduce them is to show where some ideas came from.

That implies
$$JH_+^\perp \subset H_+^\perp, \qquad JH_-^\perp \subset H_-^\perp,$$
or, equivalently, H_+ and H_- are J-invariant subspaces. It follows from the description of invariant subspaces of a selfadjoint operator that $H_\pm = \mathcal{H}_\pm$ and the decomposition is canonical. ●

4.2. Auxiliary Indefinite Hilbert Space

We are considering the Hankel operators (see (1.2))
$$\Gamma : \mathcal{H}_1 \to \mathcal{H}_2^-, \qquad \Gamma S_1 f = P_- S_2 \Gamma f,$$

Our goal is is to factorize Γ as $\Gamma = P_- T$, where T is a multiplier ($T S_1 = S_2 T$)
$$T : \mathcal{H}_1 \to \mathcal{H}_2$$

and $\|T\| \le \|\Gamma\|$. Using the idea of Ball and Helton we reduce this factorization problem to the construction of maximal nonnegative S-invariant subspace in some indefinite Hilbert space, see Section 4.3 below.

This auxiliary indefinite Hilbert space (denoted by $\mathcal{H}$) will be constructed as follows. First let us consider the Hilbert space $H = \mathcal{H}_1 \oplus \mathcal{H}_2$ with a sesquilinear form $[\,.\,,\,.\,]$ given by the operator
$$J = \begin{pmatrix} I & 0 \\ 0 & -I \end{pmatrix}.$$

If $x = \begin{pmatrix} x_1 \\ x_2 \end{pmatrix}$ and $y = \begin{pmatrix} y_1 \\ y_2 \end{pmatrix}$ then clearly

$$[x, y] = (Jx, y) = (x_1, y_1)_{\mathcal{H}_1} - (x_2, y_2)_{\mathcal{H}_2}$$

Consider the operator S in H, $S = \begin{pmatrix} S_1 & 0 \\ 0 & S_2 \end{pmatrix}$. Since $\|S_1 f\| \ge \|f\|$ and $\|S_2 g\| \le \|g\|$ for all $f \in \mathcal{H}_1$, $g \in \mathcal{H}_2$, this operator maps cone K_+ of nonnegative vectors in H into itself

$$SK_+ \subset K_+. \tag{4.2}$$

From now we assume
$$\|\Gamma\| \le 1 \tag{4.3}$$

unless the opposite is stated.

The space $\mathcal{H}$ we will be working with is given as the operator range (see [16])

$$\mathcal{H} = \begin{pmatrix} I & 0 \\ \Gamma & I \end{pmatrix} \begin{pmatrix} \mathcal{H}_1 \\ \mathcal{H}_2^+ \end{pmatrix} \subset \begin{pmatrix} \mathcal{H}_1 \\ \mathcal{H}_2 \end{pmatrix}. \tag{4.4}$$

This space is closed (because the operator $\begin{pmatrix} I & 0 \\ \Gamma & I \end{pmatrix}$ is invertible) and S-invariant. Indeed, for each $f \in \mathcal{H}_1$ and $g \in \mathcal{H}_2^+$ we have

$$\begin{pmatrix} S_1 f \\ S_2(g + \Gamma f) \end{pmatrix} = \begin{pmatrix} S_1 f \\ S_2 g + (S_2 \Gamma f - \mathbb{P}_- S_2 \Gamma f) + \mathbb{P}_- S_2 \Gamma f \end{pmatrix} =$$

$$= \begin{pmatrix} S_1 f \\ \tilde{g} + \Gamma S_1 f \end{pmatrix} = \begin{pmatrix} I & 0 \\ \Gamma & I \end{pmatrix} \begin{pmatrix} S_1 f \\ \tilde{g} \end{pmatrix},$$

where

$$\tilde{g} = S_2 g + (S_2 \Gamma f - \mathbb{P}_- S_2 \Gamma f) = S_2 g + \mathbb{P}_+ S_2 \Gamma f \in \mathcal{H}_2^+$$

The indefinite inner product of this space is induced from H, or equivalently, it is defined by the operator $A = P_{\mathcal{H}} J P_{\mathcal{H}}$. Fortunately we can find the canonical decomposition of $\mathcal{H}$, $\mathcal{H} = \mathcal{H}_+ \oplus \mathcal{H}_-$ into nonnegative and negative subspaces corresponding to the nonnegative and negative parts of spectrum of the operator A. In fact, let

$$\mathcal{H}_- = \begin{pmatrix} 0 \\ \mathcal{H}_2^+ \end{pmatrix} = \begin{pmatrix} I & 0 \\ \Gamma & I \end{pmatrix} \begin{pmatrix} 0 \\ \mathcal{H}_2^+ \end{pmatrix}$$

and

$$\mathcal{H}_+ = \left\{ \begin{pmatrix} f \\ \Gamma f \end{pmatrix} : f \in \mathcal{H}_1 \right\} = \begin{pmatrix} I & 0 \\ \Gamma & I \end{pmatrix} \begin{pmatrix} \mathcal{H}_1 \\ 0 \end{pmatrix}.$$

The subspace $\mathcal{H}_-$ is clearly negative. If $x = \begin{pmatrix} f \\ \Gamma f \end{pmatrix} \in \mathcal{H}_+$ then $[x, x] = (Jx, x) = \|f\|^2 - \|\Gamma f\|^2 \geq 0$ because $\|\Gamma\| \leq 1$, so the subspace $\mathcal{H}_+$ is nonnegative. It is clear that

$$\mathcal{H} = \mathcal{H}_+ \oplus \mathcal{H}_-, \tag{4.5}$$

($\oplus$ means usual orthogonal sum, not J-orthogonal). Let $x = x_+ \oplus x_-$, $x_\pm \in \mathcal{H}_\pm$, $x_+ = f \oplus \Gamma f$, $f \in \mathcal{H}_1$. Then

$$\begin{aligned} [x, x] &= \|f\|^2 - \|x_- + \Gamma f\|^2 = \|f\|^2 - \left(\|x_-\|^2 + \|\Gamma f\|^2 \right) = \\ &= \left(\|f\|^2 - \|\Gamma f\|^2 \right) - \|x_-\|^2 = (Jx_+, x_+) + (Jx_-, x_-) = (Ax_+, x_+) + (Ax_-, x_-), \end{aligned}$$

so by Lemma 4.1 the decomposition (4.5) is the canonical decomposition of the subspace $\mathcal{H}$. It is easy to see that the operator A_- is invertible, because it follows from the above computations that

$$A_- = \mathcal{P}_- A \mathcal{P}_- = \mathcal{P}_- J \mathcal{P}_- = I|\mathcal{H}_- .$$

Denote by $\mathcal{K}_+$ the cone of all nonnegative vectors in $\mathcal{H}$, $\mathcal{K}_+ \overset{\text{def}}{=} K_+ \cap \mathcal{H}$. Since the operator S maps nonnegative vectors into nonnegative, (see (4.2)) and the subspace $\mathcal{H}$ is S-invariant, the cone $\mathcal{K}_+$ is S-invariant too,

$$S\mathcal{K}_+ \subset \mathcal{K}_+ .$$

Remark.   Although H is a Krein space (J is invertible), $\mathcal{H}$ is, in general, not. Would we have $\|\Gamma\| < 1$ then $\mathcal{H}$ would be a Krein space as well. In principle, we can start with $\|\Gamma\| < 1$ and use the approximation to the general case $\|\Gamma\| \leq 1$. But we think that one of the advantages of working with indefinite inner product spaces instead of Krein spaces is that we are able to get the general result right away and to avoid approximation arguments.

4.3. REDUCTION OF NEHARI'S PROBLEM

In this subsection we reduce the Nehari's problem to the problem of existence of maximal S-invariant subspace in the cone $\mathcal{K}_+$ of nonnegative vectors in $\mathcal{H}$. This reduction *does not require any properties* of the operators S_1 and S_2 except trivial one that both (and so the operator S) has to be bounded. Let $\mathcal{H}$ be the indefinite inner product space described above in the previous subsection and let $S = \begin{pmatrix} S_1 & 0 \\ 0 & S_2 \end{pmatrix}$ be the shift operator in $\mathcal{H}$.

Lemma 4.2. *Let Γ be a generalized Hankel operator, $\Gamma : \mathcal{H}_1 \to \mathcal{H}_2^-$, and let $\|\Gamma\| \leq 1$. Then Γ can be factorized as*
$$\Gamma = \mathbb{P}_- T,$$
where T is a multiplier, $\|T\| \leq 1$ if and only if there exists a maximal non-negative subspace M of $\mathcal{H}$ which is S-invariant ($SM \subset M$).

Proof. Let there exist a multiplier T. Consider the subspace $M \subset H = \mathcal{H}_1 \oplus \mathcal{H}_2$
$$M = \begin{pmatrix} I & 0 \\ T & I \end{pmatrix} \begin{pmatrix} \mathcal{H}_1 \\ 0 \end{pmatrix} = \left\{ \begin{pmatrix} f \\ Tf \end{pmatrix} : f \in \mathcal{H}_1 \right\}.$$
The subspace M is S-invariant because $\begin{pmatrix} I & 0 \\ T & I \end{pmatrix}$ commutes with S and nonnegative because $\|T\| \leq 1$. To show that $M \subset \mathcal{H}$ consider a vector $\begin{pmatrix} f \\ Tf \end{pmatrix} \in M$:
$$\begin{pmatrix} f \\ Tf \end{pmatrix} = \begin{pmatrix} f \\ \mathbb{P}_- Tf + \mathbb{P}_+ Tf \end{pmatrix} = \begin{pmatrix} f \\ \Gamma f + \mathbb{P}_+ Tf \end{pmatrix} = \begin{pmatrix} I & 0 \\ \Gamma & I \end{pmatrix} \begin{pmatrix} f \\ \mathbb{P}_+ Tf \end{pmatrix} \in \mathcal{H}.$$
Let us now show that M is maximal nonnegative subspace of $\mathcal{H}$. It is a basic fact of the theory of indefinite inner product spaces (and we will prove it later, see Lemma 5.1 below) that a nonnegative subspace M is maximal if and only if $\mathcal{P}_+ M = \mathcal{H}_+$. Consider a vector $x \in M$, $x = \begin{pmatrix} f \\ Tf \end{pmatrix}$, $f \in \mathcal{H}_1$. Then clearly
$$\mathcal{P}_+ x = \begin{pmatrix} f \\ \mathbb{P}_- Tf \end{pmatrix} = \begin{pmatrix} f \\ \Gamma f \end{pmatrix},$$
so $\mathcal{P}_+ M = \mathcal{H}_+$.

Let now M be a maximal nonnegative subspace of $\mathcal{H}$. It is also a very simple fact, see again Lemma 5.1, that such a subspace can be represented in the form

$$M = \{x + Tx \ : \ x \in \mathcal{H}_+\}$$

where T is some bounded operator, $T : \mathcal{H}_+ \to \mathcal{H}_-$. We know that

$$\mathcal{H}_+ = \{ \begin{pmatrix} f \\ \Gamma f \end{pmatrix} \ : \ f \in \mathcal{H}_1\},$$

so M can be rewritten as

$$M = \{ \begin{pmatrix} f \\ Tf \end{pmatrix} \ : \ f \in \mathcal{H}_1\},$$

where T is a bounded linear operator, $T : \mathcal{H}_1 \to \mathcal{H}_2$ satisfying $\mathbb{P}_- T = \Gamma$. Since M is nonnegative subspace, we have for $x = \begin{pmatrix} f \\ Tf \end{pmatrix}$

$$[x, x] = (Ax, x) = (Jx, x) = \|f\|^2 - \|Tf\|^2 \geq 0,$$

so $\|T\| \leq 1$. Since M is S-invariant

$$S \begin{pmatrix} f \\ Tf \end{pmatrix} = \begin{pmatrix} S_1 f \\ S_2 Tf \end{pmatrix} = \begin{pmatrix} S_1 f \\ T S_1 f \end{pmatrix} \qquad \forall f \in \mathcal{H}_1,$$

and therefore $S_2 T = T S_1$. The last means that T is a multiplier. ●

4.4. THE FINAL STEP: APPLICATION OF THE IOKHVIDOV – KY FAN THEOREM

To complete the proof of Theorem 1.1 it remains now to prove the existence of the maximal S-invariant subspace M. And that will immediately follow from Theorem 3.1 above. Indeed, let A, $\mathcal{P}_\pm$, S and $\mathcal{K}_+ = \mathcal{K}_+^A$ be as defined in Section 4.2. It was shown above, see end of Section 4.2, that $S\mathcal{K}_+ \subset \mathcal{K}_+$. The subspace $\mathcal{H}_- = \begin{pmatrix} 0 \\ \mathcal{H}_2^+ \end{pmatrix}$ is clearly S-invariant. So $\mathcal{P}_+ S \mathcal{P}_- = 0$ and all assumptions of Theorem 3.1 are satisfied. Therefore, the maximal and S-invariant nonnegative subspace exists, and the theorem is proved. ●

5. Proof of the Iokhvidov – Ky Fan theorem

For the sake of completeness we present the proof of Iokhvidov – Ky Fan Theorem. The proof below is basically the proof from [20]. The only new feature is the use of Lemma 5.1 given below.

As above let A be a selfadjoint operator with invertible A_-, $K_+ = K_+^A$ be the set of all A-nonnegative vectors.

Lemma 5.1. *Assume that a subspace M is contained in K_+. Then*

1. *If $P_+ M = \mathcal{H}_+$, then there exists a unique bounded operator $T = T_M$, $T_M : \mathcal{H}_+ \to \mathcal{H}_-$ such that*
$$M = \{x + Tx \ : \ x \in \mathcal{H}_+\}$$

2. *The set $M = M_A \stackrel{\text{def}}{=} \{T_M \ : \ M \subset K_+, \ P_+ M = \mathcal{H}_+\}$ is convex, compact in the weak operator topology, subset of $L(\mathcal{H}_+, \mathcal{H}_-)$, the space of bounded linear operators from $\mathcal{H}_+$ to $\mathcal{H}_-$.*

3. *$P_+ M = \mathcal{H}_+$ if and only if it is maximal (by inclusion) subspace of K_+.*

4. *If $\operatorname{rank} A_- < \infty$ and $M \subset K_+$, then*
$$P_+ M = \mathcal{H}_+ \quad \Longleftrightarrow \quad \operatorname{codim} M = \operatorname{rank} A_-$$

Remarks. To prove 2 anb 3 we need the assumption that the operator A_- is invertible. The proof of 3 is very simple but uses the special structure of a Hilbert space. This lemma is well known and almost trivial in the case $A^2 = I$. In this case the set M is just the unit ball in $L(\mathcal{H}_+, \mathcal{H}_-)$. For the proof for the present case see [11].

To understand the main idea of the proof of the theorem, let us consider the following special case. Suppose that S maps maximal non-negative subspaces of K_+ into maximal. Then the mapping $M \mapsto SM$ induces the mapping F of M into itself,

$$F(T_M) = T_{SM}, \qquad T_M \in M. \tag{5.1}$$

It is not difficult to see that the mapping F is continuous in the weak operator topology. By Lemma 5.1 the set M is convex and compact in the weak operator topology, so we can apply Tickhonov's fixed point theorem, see [18]. By this theorem there exists a fixed point T_M of F, $F(T_M) = T_M$ or, which is equivalent, $SM = M$.

But, unfortunately, SM generally is not maximal. It is only contained in some maximal subspace. Let us express this fact in terms of corresponding angle operators. Let M be a maximal subspace of K_+, $X = T_M$ be the corresponding angle operator. Let also N be a maximal subspace of K_+ containing SM and $Y = T_N$ be its angle operator. Consider a vector $x \in M$, $x = x_+ + x_-$, $x_\pm \in \mathcal{H}_\pm$. Since $x \in M$, $x_- = X x_+$. So

$$Sx = Sx_+ + Sx_- = Sx_+ + SX x_+ .$$

On the other hand $Sx = y_+ + y_-$, $y_\pm \in \mathcal{H}_\pm$, and since $Sx \in N$ we have $y_- = Yy_+$. It is clear that

$$y_+ = P_+Sx = P_+Sx_+ + P_+SXx_+ = P_+Sx_+ + P_+SP_-Xx_+,$$

$$y_- = P_-Sx = P_-Sx_+ + P_-SXx_+ = P_-Sx_+ + P_-SP_-Xx_+.$$

Hence the equality $y_- = Yy_+$ can be rewritten as

$$P_-Sx_+ + P_-SP_-Xx_+ = Y \cdot (P_+Sx_+ + P_+SP_-Xx_+) \qquad x_+ \in \mathcal{H}_+.$$

Therefore the inclusion $M \subset N$ is equivalent to

$$P_-SP_+ + P_-SP_-X = Y \cdot (P_+SP_+ + P_+SP_-X) \tag{5.2}$$

Thus we got the following assertion. Let a function F, $F : M \times M \to L(\mathcal{H}_+, \mathcal{H}_-)$ be defined by the formula

$$F(X,Y) = Y \cdot (P_+SP_+ + P_+SP_-X) - P_-SP_+ + P_-SP_-X.$$

Then for all $X \in M$ the equation
$$F(X,Y) = 0$$

has a solution $Y \in M$. Note that since the operator P_-SP_+ is compact the function F is continuous in the weak operator topology. Therefore Theorem 3.1 follows from the lemma below. This lemma is due to I.S. Iokhvidov [20] and says that the multivalued function f defined by F (f maps $X \in M$ to the non-empty set of all $Y \in M$ such that $F(X,Y) = 0$) has a "fixed point". The case $F(X,Y) = Y - \phi(X)$ where ϕ is single valued mapping gives rise to Tyckhonov's fixed point theorem.

Lemma 5.2. *Let L be a Hausdorff locally convex linear topological space and $\mathcal{X}$ be a convex compact subset of L. Let $F : \mathcal{X} \times \mathcal{X} \to L$ be a continuous mapping satisfying the conditions below:*

1. $F(x, \alpha y_1 + (1 - \alpha)y_2) = \alpha F(x, y_1) + (1 - \alpha)F(x, y_2), \qquad 0 \le \alpha \le 1;$

2. $\forall x \in \mathcal{X} \ \exists y \in \mathcal{X} : \quad F(x, y) = 0.$

Then there exists $x_0 \in \mathcal{X}$ such that

$$F(x_0, x_0) = 0.$$

The proof of the lemma is based on the following result of Ky Fan [21]:

Lemma 5.3. *Let $\mathcal{X}$ be a convex, compact subset in a Hausdorff linear topological space L and let U be a closed subset in $\mathcal{X} \times \mathcal{X}$ such that*

1. $(x, x) \in U \qquad \forall x \in \mathcal{X};$

2. *For all $x \in \mathcal{X}$ the set $\{y \in \mathcal{X} \; : \; (x,y) \notin U\}$ is convex (or empty).*

Then there exists $x_0 \in \mathcal{X}$ such that $(x_0, y) \in U$ for all $y \in \mathcal{X}$.

Proof of Lemma 5.2.   Let $\{p_\nu\}$ be a collection of semi-norms defining the topology on L. Let

$$\xi_\nu \stackrel{\text{def}}{=} \{x \in \mathcal{X} \; : \; p_\nu(F(x,x)) = 0\}.$$

It follows from the compactness of $\mathcal{X}$ that to prove the lemma it is sufficient to show that for any finite collection $\nu_1, \nu_2, ..., \nu_n$ the intersection $\cap_{k=1}^n \xi_{\nu_k}$ is non-empty. Let us prove this.

Fix a collection $\nu_1, \nu_2, ..., \nu_n$ and set

$$U = \{(x,y) \in \mathcal{X} \times \mathcal{X} \; : \; \sum_{k=1}^n p_{\nu_k}(F(x,x)) \le \sum_{k=1}^n p_{\nu_k}(F(x,y))\}.$$

It is easy to see that the set U satisfies all the assumptions of Lemma 5.3. Therefore there exists $x_0 \in \mathcal{X}$ such that $(x_0, y) \in U$ for all y in $\mathcal{X}$. According to the assumption 2 of Lemma 5.2 there exists a point $y_0 \in \mathcal{X}$ such that $F(x_0, y_0) = 0$. But according to the definition of U we have

$$\sum_{k=1}^n p_{\nu_k}(F(x_0, x_0)) \le \sum_{k=1}^n p_{\nu_k}(F(x_0, y_0))$$

and, therefore $p_{\nu_k}(F(x_0, x_0)) = 0$, $1 \le k \le n$. Thus the intersection $\cap_{k=1}^n \xi_{\nu_k}$ contains the point x_0 and so is non-empty. The lemma is proved.

To prove Lemma 5.3 (and so the theorem) we need the following fact from topology. The symbol $\operatorname{convh}(E)$ denotes here and below the convex hull of the set E.

Lemma 5.4. (Knaster – Kuratowski – Mazurkiewicz, [24]) *Let $t_1, t_2, ..., t_n \in \mathbb{R}^{n-1}$ and let $\operatorname{convh}(t_1, t_2, ..., t_n)$ be an $(n-1)$-dimensional simplex (i.e. the points $t_1, t_2, ..., t_n$ are not contained in any proper subspace of $\mathbb{R}^{n-1}$). Let $V_1, V_2, ..., V_n$ be a collection of open subsets of the simplex such that for any subcollection $\sigma \subset \{1, 2, ..., n\}$ of indices*

$$(\cup_{k \in \sigma} V_k) \cap \operatorname{convh}(t_k \; : \; k \in \sigma) = \varnothing. \qquad (5.3)$$

Then the sets $V_1, V_2, ..., V_n$ do not cover the simplex.

Proof of Lemma 5.3. Let U_y be the section of the set U,

$$U_y \stackrel{\text{def}}{=} \{x \in \mathcal{X} \; : \; (x,y) \in U\}.$$

To prove the lemma we have to prove that $\cap_{y \in \mathcal{X}} U_y \ne \varnothing$. By virtue of the compactness of $\mathcal{X}$, it is sufficient to show that for any finite collection $y_1, y_2, ..., y_n$, $y_k \in \mathcal{X}$, we have $\cap_{k=1}^n U_{y_k} \ne \varnothing$ or, equivalently $\cup_{k=1}^n V_{y_k} \ne \mathcal{X}$, where $V_{y_k} = \mathcal{X} \setminus U_{y_k}$.

The sets U_{y_k} are closed, so V_{y_k} are open. It is easy to see that

$$(\cup_{k \in \sigma} V_{y_k}) \cap \operatorname{convh}(y_k \; : \; k \in \sigma) = \varnothing.$$

Indeed, if x in convh $(y_k \ : \ k \in \sigma)$ belongs to the intersection $\cup_{k \in \sigma} V_{y_k}$ then clearly

$$y_k \in \{y \in \mathcal{X} \ : \ (x, y) \notin U\}, \qquad k \in \sigma. \tag{5.4}$$

Since the set $\{y \in \mathcal{X} \ : \ (x, y) \notin U\}$ is convex (assumption 2 of the lemma) it follows from (5.4) that $(x, x) \notin U$ that contradicts the assumption 1 of the lemma.

Therefore, if convh $(t_1, t_2, ..., t_n)$ is an $(n-1)$-dimensional simplex in $\mathbb{R}^{n-1}$, τ is the affine mapping of this simplex into convh $(y_1, y_2, ..., y_n)$, $\tau(t_k) = y_k$ then the sets $V_k \overset{\text{def}}{=} \tau^{-1}(V_{y_k})$ satisfy the assumptions of Lemma 5.4. Consequently $\bigcup_{k=1}^{n} V_k \neq$ convh $(t_1, t_2, ..., t_n)$ and thus also

$$\text{convh} \, (y_1, y_2, ..., y_n) \not\subset \bigcup_{k=1}^{n} V_{y_k}.$$

REFERENCES

[1] Z. NEHARI, On bounded bilinear forms. *Ann. of Math.*, **65**(1957), 153-162.

[2] C. FOIAS AND A. E. FRAZHO, "The commutant lifting approach to interpolation problems," Operator Theory Advances and Applications, **42**, Birkhauser, Basel – Boston, 1990.

[3] S. C. POWER, "Hankel operators on Hilbert space," Research notes in Math., **64**, Pitman Adv. Publ. Progr., Boston etc., 1982.

[4] V. M. ADAMYAN, D. Z. AROV, AND M. G. KREIN, Analytic properties of Schmidt pairs of Hankel operators and the generalized Schur – Takagi problem, *Matem. sbornik* **86**(1971), N 1, 33-73 (Russian)

[5] V. M. ADAMYAN, D. Z. AROV, AND M. G. KREIN, Infinite block Hankel matrices and some related continuation problems. *Izv. Akad. Nauk Armjan. SSR Ser. Mat.* **6**(1971), 87-112 (Russian)

[6] R. AROCENA, M. COTLAR AND C. SADOSKY, Weighted inequalities in L^2 and lifting property. *in* Math. Anal. and Appl., Adv. in Math. Suppl. Stud., **7A**(1981), 95-128.

[7] R. AROCENA, AND M. COTLAR, A generalized Herglotz – Bochner theorem and L^2-weighted inequalities with finite measures. *in* Proc. Conf. Harm. Anal. in honor of A. Zygmund, Chicago 1981.

[8] R. AROCENA, AND M. COTLAR, Dilation of generalized Toeplitz kernels and some vectorial moment and vector problems, *in* "Lect. Notes Math.," **908**(1982).

[9] M. COTLAR AND C. SADOSKY, Weighted and two-dimensional Adamyan – Arov – Krein theorems and analogues for Sarason commutators, *Institut Mittag-Leffler*, *Report* No 24, 1990/1991 (to appear in Integral Equations ant Operator Theory).

[10] J. A. BALL AND J. W. HELTON, A Beurling – Lax theorem for the Lie group $U(m, n)$ which contains most classical interpolation theory. *J. Oper. Theory* **9**(1983), 107-142.

[11] S. R. TREIL, Adamyan – Arov – Krein theorem: vector valued version, *Zap. Nauchn. Semin. LOMI* **141**(1985), 56-81 (Russian).

[12] C. FOIAS AND A. TANNENBAUM, On the singular values of the four-block operator and certain generalized interpolation problems, *in* "Analysis and Partial Differential Equations, Lect. Notes in Pure Math.," **122**, 483-493, Decker, NY, 1990.

[13] J.A. BALL AND E.A. JONCKHEERE, The four-block Adamjan – Arov – Krein problem, *J. Math. Anal. Appl.* **170** (1992), 322-342.

[14] K. GLOVER, D. LIMEBEER AND Y.S. HUNG, A structured approximation problem with application to frequency weighted model reduction, *IEEE Trans. Automatic Control* **37** (1992), 447-465.

[15] N. K. NIKOLSKII, "Treatise of the shift operator," Springer-Verlag, NY etc. 1985.

[16] J. W. HELTON, "Operator theory, analytic functions, matrices, and electrical engineering," Conf. Board of Math. Sci. Conf. Series in Math., **68**(1986).

[17] B. A. FRANCIS, "A course of H^∞ control theory," Lect. Notes in Control and Information Sci., **88**, 1987, Spriner-Verlag.

[18] N. DANFORD AND L. SCHWARTZ, "Linear operators. Part I:General theory," New York, 1957.

[19] B. SZ.-NAGY AND C. FOIAS, "Harmonic analysis of operators on Hilbert space," Akadémiai Kiadó, Budapest, 1970.

[20] I. S. IOKHVIDOV, On a lemma of K. Fan generalizing Tyckhonov's fixed point principle, *Dokl Akad Nauk SSSR* **159**(1964), 501-504 (Russian).

[21] K. FAN, A generalization of Tyckhonov's fixed point theorem, *Math. Ann.*, **142**(1961), 305-310.

[22] T. YA. AZIZOV AND I. S. IOKHVIDOV, "Linear operators in spaces with indefinite metric and their applications," John Willey & Sons, New York etc., 1989.

[23] J. BOGNAR, "Indefinite inner product spaces," Springer-Verlag, 1974.

[24] B. KNASTER, C. KURATOWSKI AND S. MAZURKIEWICZ, Ein Beweis des Fixpunktsatzes fur n-dimensionale Simplexe, *Fund. Math.* **14**(1929), 132-137.

Sergei Treil
Department of Mathematics,
Michigan State University
East Lansing, MI 48824
e-mail: treil@math.msu.edu

Alexander Volberg
Department of Mathematics,
Michigan State University
East Lansing, MI 48824
e-mail: volberg@math.msu.edu

Mathematical Subject Classification: 47B35, 47A45 — primary
                                          47H10 — secondary

Titles previously published in the series

OPERATOR THEORY: ADVANCES AND APPLICATIONS
BIRKHÄUSER VERLAG

52. **S. Prössdorf, B. Silbermann:** Numerical Analysis for Integral and Related Operator Equations, 1991, (3-7643-2620-4)

53. **I. Gohberg, N. Krupnik:** One-Dimensional Linear Singular Integral Equations, Volume I, Introduction, 1992, (3-7643-2584-4)

54. **I. Gohberg, N. Krupnik:** One-Dimensional Linear Singular Integral Equations, Volume II, General Theory and Applications, 1992, (3-7643-2796-0)

55. **R.R. Akhmerov, M.I. Kamenskii, A.S. Potapov, A.E. Rodkina, B.N. Sadovskii:** Measures of Noncompactness and Condensing Operators, 1992, (3-7643-2716-2)

56. **I. Gohberg** (Ed.): Time-Variant Systems and Interpolation, 1992, (3-7643-2738-3)

57. **M. Demuth, B. Gramsch, B.W. Schulze** (Eds.): Operator Calculus and Spectral Theory, 1992, (3-7643-2792-8)

58. **I. Gohberg** (Ed.): Continuous and Discrete Fourier Transforms, Extension Problems and Wiener-Hopf Equations, 1992, (3-7643-2809-6)

59. **T. Ando, I. Gohberg** (Eds.): Operator Theory and Complex Analysis, 1992, (3-7643-2824-X)

60. **P.A. Kuchment:** Floquet Theory for Partial Differential Equations, 1993, (3-7643-2901-7)

61. **A. Gheondea, D. Timotin, F.-H. Vasilescu** (Eds.): Operator Extensions, Interpolation of Functions and Related Topics, 1993, (3-7643-2902-5)

62. **T. Furuta, I. Gohberg, T. Nakazi** (Eds.): Contributions to Operator Theory and its Applications. The Tsuyoshi Ando Anniversary Volume, 1993, (3-7643-2928-9)

63. **I. Gohberg, S. Goldberg, M.A. Kaashoek:** Classes of Linear Operators, Volume 2, 1993, (3-7643-2944-0)

64. **I. Gohberg** (Ed.): New Aspects in Interpolation and Completion Theories, 1993, (3-7643-2948-3)

65. **M.M. Djrbashian:** Harmonic Analysis and Boundary Value Problems in the Complex Domain, 1993, (3-7643-2855-X)

66. **V. Khatskevich, D. Shoiykhet:** Differentiable Operators and Nonlinear Equations, 1993, (3-7643-2929-7)

67. **N.V. Govorov** †: Riemann's Boundary Problem with Infinite Index, 1994, (3-7643-2999-8)

68. **A. Halanay, V. Ionescu:** Time-Varying Discrete Linear Systems Input-Output Operators. Riccati Equations. Disturbance Attenuation, 1994, (3-7643-5012-1)

69. **A. Ashyralyev, P.E. Sobolevskii:** Well-Posedness of Parabolic Difference Equations, 1994, (3-7643-5024-5)

70. **M. Demuth, P. Exner, G. Neidhardt, V. Zagrebnov** (Eds): Mathematical Results in Quantum Mechanics. International Conference in Blossin (Germany), May 17-21, 1993, 1994, (3-7643-5025-3)

71. **E.L. Basor, I. Gohberg** (Eds): Toeplitz Operators and Related Topics. The Harold Widom Anniversary Volume. Workshop on Toeplitz and Wiener-Hopf Operators, Santa Cruz, California, September 20–22, 1992, 1994 (3-7643-5068-7)